VGM Opportunities Series

OPPORTUNITIES IN
ENVIRONMENTAL
CAREERS

Odom Fanning

Foreword by
Jay D. Hair, Ph.D.
President
National Wildlife Federation

VGM Career Horizons
a division of *NTC Publishing Group*
Lincolnwood, Illinois USA

Cover Photo Credits
Clockwise from upper left: G.D. Searle and Company; USDA Soil Conservation Service; USDA Soil Conservation Service; University of Wisconsin-Stevens Point (Doug Moore photo).

Library of Congress Cataloging-in-Publication Data
Fanning, Odom.
   Opportunities in environmental careers / Odom Fanning ; foreword by Jay D. Hair. — 6th ed.
      p.   cm. — (VGM opportunities series)
   Includes bibliographical references.
   ISBN 0-8442-4583-6 (h). — ISBN 0-8442-4584-4 (p).
   1. Environmentalists—Vocational guidance—United States.   2. Environmental policy—Vocational guidance—United States.   3. Environmental education—Vocational guidance—United States.   4. Environmental sciences—Vocational guidance—United States.   5. Environmental engineering—Vocational guidance.
I. Title.   II. Series.
GE60.F36   1995
363.7'0023'73—dc20   95-21865
                                                          CIP

Published by VGM Career Horizons, a division of NTC Publishing Group
4255 West Touhy Avenue
Lincolnwood (Chicago), Illinois 60646-1975, U.S.A.
© 1996 by NTC Publishing Group. All rights reserved.
No part of this book may be reproduced, stored in a retrieval
system, or transmitted in any form or by any means,
electronic, mechanical, photocopying, recording or otherwise,
without the prior permission of NTC Publishing Group.
Manufactured in the United States of America.

5 6 7 8 9 0 VP 9 8 7 6 5 4 3 2 1

DEDICATION

This edition is dedicated to Sarah Elaine and to Michael,
to all of their generation,
and to its environment.

CONTENTS

Catalysts for environmental change. The new environmentalism.
The chain of "I's." What questions should you be able to answer
after reading this book?

Organizations offering guidance on environmental education and
careers. Liberal arts colleges offering specialized environmental
curricula. U.S. universities offering interdisciplinary environmental
curricula. Canadian universities offering diversified environmental
curricula. Where should you write for more information?

Professions covered. Biology. Ecology. Environmental chemistry
and chemical engineering. Where should you write for more
information? Subscription newsletters listing current openings in
all environmental fields. Guides to job search.

Professions covered. "Healthy People 2000." Professionalism.
Environmental engineering. Environmental health sciences.
Environmental medicine. The environmental physician.

Public health veterinary medicine. Environmental sanitation. Industrial hygiene. Where should you write for more information?

FOREWORD

As we stand on the threshold of a new millennium, we are grateful to the conservationists of the past and awed by our obligations to the generations ahead. Thankful as we are for the resources of North America, we must remain ever vigilant so that they are used wisely and not squandered wantonly.

Our commitment to a sustainable future begins with a strong conservation ethic among our young persons, an ethic that must be nurtured by example and furthered through education. Thus committed, we can look forward to the twenty-first century as The Century of the Environment.

More and more of the vital activity in managing our open spaces, air and water, wildlife, and habitats is occurring at local and regional levels across the country, and fewer key policy decisions are being made by the U.S. Congress and carried out by federal agencies in Washington, D.C. This is proper, but federal legislation is still vital. High on most environmentalists' priority lists, as this is written, are reauthorization of several of the basic pollution control laws, important changes in the 1872 Mining Law, and the laws governing livestock grazing on western public lands. And there are the ongoing battles to save endangered species from extinction, and to preserve precious old-growth forests and wetlands.

These are not partisan issues. Sound environmental stewardship is a value shared by Americans from all walks of life and of every affiliation. Further, the majority of Americans recognize that there need be no conflict in promoting economic growth, protecting human health, minimizing pollution hazards, and safeguarding natural resources for our children and grandchildren so they can enjoy life to the fullest and prosper as we have.

Perhaps the attribute that has distinguished the National Wildlife Federation over its six decades is a history of reaching out from one generation to nurture generations to come. Since its founding in 1936, the federation has been a good environmental steward, involving millions of citizens in this cause. While maintaining high-quality public and professional education and outdoor recreation programs, we have in recent years expanded our activities to new, more culturally diverse communities in the nation's urban centers. And we are harnessing young adults' creative energies; for example, providing environmentally sound experiences to students through another of our newer programs, the Campus Ecology program.

I am privileged once again to contribute a foreword to this sixth, revised edition of VGM Career Horizons' *Opportunities in Environmental Careers.* In his research, Odom Fanning has found a continuing strong desire among young persons for education in environmental arts and sciences, and a continuing response from colleges and universities in providing a range of environmental training and studies. Moreover, the Bureau of Labor Statistics projects a healthy demand into the twenty-first century for well-educated personnel in many areas of environmental management.

Polls consistently show that a majority of Americans support well-thought-out environmental actions and essential expenditures to maintain a sustainable future. Such a future includes an environmentally literate society, a well-prepared work force, a strong economy that provides good jobs and careers for all persons, and a continuing commitment to protecting our environment through careful use of its wonderfully rich resources.

Jay D. Hair, Ph.D.
President
National Wildlife Federation
Washington, D.C.

PREFACE

This sixth edition of *Opportunities in Environmental Careers* is the first to look ahead to a new decade in a new century in a new millenium. It recognizes the changes wrought by environmentalism in the last third of the twentieth century And it analyzes the implications for young persons contemplating their career horizons in the first third of the twenty-first century.

Environmentalism is in the public interest; strong evidence confirms that a majority of Americans hold that conviction. Yet, inevitably, there are opposing stands on environmental issues, and cost-benefit analyses favor one adversary or another at one time or another. It is assumed in this book that what is called the *new environmentalism* has established itself over the last thirty years, and over the next thirty will be stronger than ever in three components: education, science, and management.

The new environmentalism employs its resources in new ways. It is ends-oriented, not means-oriented. It acknowledges, as does the foreword of Dr. Jay Hair, the present generation's debts "to the conservationists of the past . . .[and] our obligations to the generations ahead."

This book's focus is on the end objective of environmental careers for the present generation of young people, and on what the immediate past portends for their immediate future—and for our country's future.

The Bureau of Labor Statistics has advanced a theory about the significance of generational turnover, especially in a rapidly growing field. Here is the way it works in environmental management: Within a few years of the first Earth Day, the older generation of environmentalists— environmental managers or professionals of the 1960s—retired and was replaced by a new generation. By the mid-1990s, those environmental-ists were retiring, to be replaced by the second generation of *new environmentalists* (a term to be defined in Chapter 1). Each generation has been larger than the one before; employment has grown from an estimated half million to one million, perhaps as large now as two million. A doubling every decade cannot long continue, but even with annual growth slowed to a low, single-digit percentage, the field of environmental management is sure to remain strong and viable well into the twenty-first century.

The decade 2001–2010 should see a third generation of new environmentalists assuming responsible positions in the work force. Hence, for three reasons, this edition of *Opportunities in Environmental Careers* is directed precisely toward this emerging third generation of aspiring environmentalists. The first reason is to introduce the many career opportunities that lie ahead.

The second reason is to explain a new concept in environmental management, the chain of "I's," which will be such a determinant of their future.

And a third purpose is to describe the education necessary to qualify for an environmental career. The emphasis is on science and engineering as the principal tools for making environmental change. Beyond the purview of this book are the many arts which also are necessary, such as economics, history, journalism, mediation, population, and statistics.

For each of a score of major environmental professions, this book examines such questions as

- What education would you need to prepare for such a career?
- Where would you work?
- What would you earn?
- What is the job's future?

The Council on Environmental Quality is optimistic, and so should we all be. It speaks a simple truth when it says:

Americans believe strongly that environmfental quality is an essential component of their long-term health and economic prosperity. Americans have demonstrated that they have the will to protect environmental quality and the capacity to act with energy, creativity, and a deep-seated sense of responsibility for future generations.

Odom Fanning
Bethesda, Maryland

ABOUT THE AUTHOR

Odom Fanning is a veteran science writer and accredited Washington correspondent who has won numerous awards for his environmental and medical writing.

A graduate of Emory University, Fanning was a science writer for the *Atlanta Journal* before becoming the first information officer for the Centers for Disease Control and Prevention (CDC); then research information director for the Georgia Institute of Technology in Atlanta and, later, CBS Broadcast Laboratories in New York. Since moving to the nation's capital, he has managed national programs in environmental health, occupational health and safety, and energy conservation; and on three occasions he has been assigned to the White House to edit presidential reports.

When Congress created the White House Council on Environmental Quality, Fanning was invited to become editor-in-chief of *The President's First Annual Report on Environmental Quality,* 1970. Following the first Earth Day, he wrote numerous magazine articles on environmental topics, this first career guide to environmental careers, the first college textbook on environmental citizen action, and a handbook on environmental management for developing countries.

In 1976—citing such contributions, particularly "his hallmark book *Opportunities in Environmental Careers"*—the North American Association for Environmental Education awarded Fanning its Citation of Merit for Outstanding Service to Environmental Education.

CHAPTER 1

THE NEW ENVIRONMENTALISM

> The Congress recognizes that each person should enjoy a healthful environment and that each person has a responsibility to contribute to the preservation and enhancement of the environment.—The National Environmental Policy Act of 1969.

Young persons today have known no other society than the one into which they were born, and may take for granted what came before. They may not care much about tradition. Even the social revolutions of the 1960s and 1970s, which so drastically changed our society, may seem distant when one watches documentaries of the period.

Our environmental heritage, like our social heritage, was established over many decades, by all citizens, but especially by conservationists. One of the driving forces of the human rights movements of the 1960s was a fierce determination to establish permanent protection of many newly recognized environmental and consumer rights. The public interest law movement, epitomized by Ralph Nader, tied these two forces together; coalitions were formed combining environmental and consumer rights forces with other groups, such as those promoting civil rights.

Many young Americans were in revolt against misuse of government authority and particularly in turmoil over the devastation that was being inflicted on the American environment. The impact of the war in southeast Asia, and especially the effects of television coverage during this period, cannot be underestimated: people watched actual occurrences in Vietnam and were repelled by what they saw.

CATALYSTS FOR ENVIRONMENTAL CHANGE

Silent Spring, a book by Rachel Carson published in 1962, dramatized the price society pays for the indiscriminate use of technological advances. Carson's thesis was that "we have allowed these chemicals [pesticides] to be used with little or no investigation of their effect on soil, water, wildlife, and man himself. Future generations are unlikely to condone our lack of prudent concern for the integrity of the natural world that supports all life." *Silent Spring* galvanized public opinion for the environmental changes to come.

Young persons, in particular, became aroused over pollution, blight, deterioration, waste, and destruction of our natural resources, and over the terrible hurt to everybody's health. Late in the 1960s, the media began to catalyze public opinion behind a change in social and political policies. Then, on January 28, 1969, a gigantic oil spill occurred offshore at Santa Barbara, California. The television pictures of seabirds, their feathers coated with oily gunk, suffering and dying while people who cared tried to rescue them, symbolized all forms of technology's overuse in conflict with the environment. Capturing the national attention, that single event provoked the public into action.

The public health movement was two and one-half centuries old, and the conservation movement one century old, at the time. Environmental stewardship was entrenched in our American value system. But most observers credit that catastrophe on the California coast with marking the start of the *new* environmental era. Modest organization work to promote the cause of the environment continued for the next year. Then, the following April of 1970, an estimated twenty million Americans participated in the first Earth Day, a more-or-less spontaneous, unexpectedly successful celebration.

Concurrently, a great deal of political groundwork was being laid. Through a rare combination of imagination, political leadership, and chance, the Congress produced a gem of a law in the National Environmental Policy Act of 1969 (NEPA). It was full enough of rhetoric to satisfy the most ardent Earth Day demonstrator. Every law has its rhetoric, but NEPA had more: a strong action-forcing device. Its Section 102(2)(c) requires an environmental impact statement in advance of any federal action that might significantly affect the quality of the environment. This requirement has revolutionized U.S., even international, decision making. It rhetorically asks governmental agencies: "Wait! Do we know all we should know about what you propose to do? Do we understand its consequences? Can we make it safer for people and more benign for the environment? Do the benefits outweigh the costs?"

If such questions had been asked—and if the will and the technology had existed to answer them—Los Angeles might have been built so as to minimize its smog and traffic congestion. The automobile might have been developed so as to avoid harmful emissions. Certainly hazardous and toxic chemicals would not have been dumped in thousands of locations, from which they definitely cause illness, and probably cause much of the rising incidence of cancer, miscarriages, birth defects, mental retardation, and other serious threats to health and life.

We may not be able to do much about some of the environmentally related threats to our health, but we can unburden future generations by not making similar mistakes in our own uses of technology. The major social movement which offers the potential for stepping back, taking a second look, and making more rational social, political, and technological decisions is what we call the *new environmentalism.*

THE NEW ENVIRONMENTALISM

The new environmentalism has three facets: education, science, and management. It employs them in new ways. It is ends-oriented, not means-oriented. Contrast the old with the new as follows:

Education. It might seem that defining environmental education would be as simple and as straightforward as defining medical education or engineering education, but such is not the case. A major problem in defining environmental education, compared to other types of education, is that it is all-encompassing. The difficulty begins with the environment's not being a discipline. Medical education is education *in* medicine. Engineering education is education *in* engineering, and so forth throughout scores of disciplines. But environmental education is education *about* the environment.

Science. The new environmentalism also embraces environmental science. Traditionally, scientists and engineers worked on narrow disciplinary problems such as those having to do with climate, air turbulence, estuaries, forests, epidemics, earthquakes, and groundwater, to name a few. After 1970, the perspective broadened in response to governmental support for interdisciplinary research and development.

There were other reasons for this broadening, too. New technology such as high-speed computers relieved scientists of much of the drudgery which long had characterized their work. Machines could process vast amounts of data with the speed of light, making it possible to handle this information explosion. Scientists could devote their time to things that were more "fun," one of which was interacting with fellow scientists in a teamwork approach to more broad-based environmental problems. For the first time, they could think in terms of "managing" or manipulating environmental phenomena, as easily as they could conduct scientific experiments in the laboratory.

Management. Environmental management is an action term; it refers to all activities, public and private, undertaken to achieve the goals of environmental quality. In such activities lie many jobs and careers. This is the author's definition:

> Environmental management is an interdisciplinary, integrated effort, involved with the very fabric of people's lives, focused on interrelated environmental problems and employing the findings of science, the techniques of engineering, and the understanding of the social sciences to preserve the human environment, to utilize natural resources, and to better society.

THE CHAIN OF "I'S"

Note, in the definition, the "chain of I's": *interdisciplinary/integrated/involved.* Take each link:

Interdisciplinary. Whereas once it was thought sufficient for disciplines to stand alone, the sciences—especially the natural sciences and their forms of application as in engineering and technology—now are recognized as complex and closely interconnected. Biology, chemistry, ecology, physics, and other scientific disciplines, linked as in a chain, must be employed together in the solution of complex environmental problems. That is why we speak of many fields, including environmental management, as interdisciplinary.

Integrated. Environmental management is integrated—parts incorporated into a larger whole. To the sciences must be added such other disciplines as engineering, business management and public administration, communications, economics, education, history, the humanities, law, philosophy, political science, psychology, sales and marketing, and the social sciences. Environmental management is an integrated form of interdisciplinary endeavor.

Involved. Environmental management endeavors, while interdisciplinary and integrated, must be undertaken with the full appreciation that they are focused on the human environment, therefore people are affected. For this reason, environmental managers must be involved in the broader society; they must be good environmental citizens, whether or not their profession has a code of ethics, requires a license, or confers a title such as Doctor. Some of the ways in which this citizen involvement occurs are to be found throughout the chapters to follow.

WHAT QUESTIONS SHOULD YOU BE ABLE TO ANSWER AFTER READING THIS BOOK?

1. What type of college or university would you prefer to attend, and are you likely to find a suitable environmental studies curriculum there?
2. What are some environmental fields you could enter with the bachelor's degree?
3. Biotechnology is an emerging area of the biological sciences. What disciplines contribute to biotechnology? What are some of its applications, and why is it a viable career option?
4. What is ecology, and what do ecologists do?
5. What is the significance of conservation biology?

6. Why does one professor call a degree in environmental chemistry "a ticket to a career in business, sales, journalism, or law"?
7. What are the advantages to alternate periods of work and college study?
8. What are some of the opportunities for environmental engineers overseas?
9. What are some of the national health objectives for the year 2000? How does success in meeting them depend upon adequate supplies of environmental health scientists and other workers? What kinds of workers are needed?
10. Why do some conservation fields now substitute the noun *management* for *conservationist?*
11. Why is agricultural science considered an environmental field?
12. Why do the states grant licenses to some environmental professionals and not to others?

CHAPTER 2

EDUCATION FOR THE NEW ENVIRONMENTALISM

Environmental education is a [process that] promotes the analysis and understanding of environmental issues and questions as the basis for effective education, problem solving, policy making, and management. [The purposes of environmental education are] to foster the education of skilled individuals able to understand environmental problems and possessing the expertise to devise effective solutions to them; and the development of a citizenry conscious of the scope and complexity of current and emerging environmental problems, and supportive of solutions and polices which are ecologically sound.—The preamble to the Constitution of the North American Association for Environmental Education, 1994.

If you are college-bound, your choice of college may be determined by finances, geographic preference, grades and SATs, the institution's reputation, its rating in various directories, how much scholarship funding it has available, and other factors, subjective as well as objective (including perhaps where your parents went to college).

It is unwise, even for the student highly motivated toward an environmental career, to focus too early on only one institution or type. You can get a good grounding in science and the liberal arts at almost any of the 1,200 community colleges or 1,800 four-year colleges and universities in America. Find a school at which you feel comfortable but challenged. Many students transfer, especially between sophomore and junior years, so an initial choice can easily be changed, usually without significant loss of credits.

This chapter focuses on environmental science/environmental studies (sometimes abbreviated ES/ES). This is a type of higher education that is *interdisciplinary/integrated/involved.* A bachelor's degree in ES may qualify you for entry into some fields for which the bachelor's degree is standard for professional practice, e.g., agricultural sciences, fisheries, soils or wildlife conservation, and range management. Even in the field of environmental health, there are some professions, e.g., industrial hygiene, and several categorical specialties in which a postgraduate degree may not be necessary.

Furthermore, there are new functions which the ES/ES graduate may be better qualified to perform than someone who has a narrow science or engineering specialty. An example may be preparing environmental impact studies and statements or editing and publishing scientific, engineering, or economic reports. A good grounding in English and social science, beneficial to any professional, would be especially so in this work.

ORGANIZATIONS OFFERING GUIDANCE ON ENVIRONMENTAL EDUCATION AND CAREERS

North American Association for Environmental Education

The North American Association for Environmental Education (NAAEE), established in 1971, is a professional membership organization with more than 2,500 members in the United States and over 40 other countries. It has significant membership in Canada and Mexico, has met in both countries, and is equally attentive to their environments and to that of the United States. In fact, it is concerned with the social and environmental betterment of people everywhere, through environmental education.

"NAAEE is made up of people who have thought seriously—over lifetimes—about how people become literate concerning environmental issues," says Edward McCrea, the association's executive director. "Association members believe education must go beyond consciousness raising. Environmental education must prepare people to think about the difficult decisions they must make concerning environmental steward-ship and [how] to work together towards the resolution of environmental problems."

Environmental education must integrate knowledge from the natural sciences, social sciences, and humanities, according to NAAEE's ten guiding principles, but information and analysis are only part of an effective education program. McCrea explains: "NAAEE believes that to be truly effective this body of knowledge must be integrated into all aspects of the curriculum and must be viewed within a broad social context. Environmental education must build on a foundation of relevancy, and employ education techniques for reaching the widest array of audiences."

The NAAEE Board of Directors has voted to take the lead in developing, by consensus, a model set of voluntary national standards for environmental education (EE). Areas to be covered are EE materials, EE student performance, and environmental educator performance standards.

Chairing the effort is Prof. Bora Simmons of Northern Illinois University at DeKalb, who is quoted in the *Environmental Communicator* as saying: "We believe that environmental education should be an integral part of every student's schooling. Infused throughout the curriculum, EE supports the high standards set by the traditional disciplines. More importantly, environmental education encompasses the knowledge, skills, and inclinations that are essential to maintaining an equilibrium between quality of life and quality of the environment."

Self-descriptive are three of NAAEE's four membership sections: college and university environmental programs, conservation education, and elementary and secondary education. A fourth, nonformal, is for those who work in museums, zoos, camps, nature centers, and youth organizations. (A committee on environmental health may lead to a fifth section.)

"The field of environmental studies has seen tremendous growth over the past five or more years," writes College and University Section Chair Jim Elder, who is at The School for Field Studies, Beverly, Massachusetts. "Rough estimates are that the number of undergraduate ES programs has more than doubled in the United States, and overall enrollment in ES programs has gone up more than fourfold during this period."

NAAEE offers student membership at a reduced rate. Membership in a section is at no additional charge, as are access to a skills bank, jobs and internships listings, and other services. All members receive the bi-monthly *Environmental Communicator* and may attend the annual conference. NAAEE publications, and environmental periodicals issued by other organizations, are available to NAAEE members at reduced rates. (For the address of NAAEE and other organizations described here, see listings at the end of this chapter.)

The Environmental Careers Organization

The Environmental Careers Organization (ECO) is a national, non-profit—but not membership—organization dedicated to the development of individuals' environmental careers. "This is a lifelong learning activity where you live, work, and play—through your job, philanthropy, and volunteerism," states ECO's latest annual report. "Our mission is to protect and enhance the environment through the development of professionals, the promotion of careers, and the inspiration of individual action."

ECO, founded in 1972 and headquartered in Boston, has regional offices there and in Cleveland, Seattle, San Francisco, and Tampa. Corporations, foundations, and individuals contribute to its multimillion dollar annual budget. ECO has an "alumni network" numbering over 6,000.

"The development of the environmental profession can be compared with the historical development of other professions," says John R. Cook, Jr., ECO's founder and president. "Right now, I believe the field is going through a maturing phase, just as medicine or law did hundreds of years ago. First, everyone is off doing what works best for [him or her]. Gradually, knowledge and experience grow and are shared; this is specialization and standardization, accreditation, and development of professional ethics and codes of conduct. It is very exciting to be in a profession during this formative period, and it gives individuals the opportunity to have a profound effect on the field for generations to come."

Under Cook's leadership, ECO has become a large, successful organization. Its Environmental Placement Services offers short-term associate positions for college students and recent graduates, as many as 500 per year. They are employed on assignments in federal, state, and local agencies, corporations, and nonprofit organizations, working an average of 24 weeks at median salaries ranging from $300 to $700 per week. Cook says that nearly all of the thousands of ECO associates of the past two decades are still working in the environmental field.

The Diversity Initiative strives to increase the presence and numbers of people of color working in the environmental professions. ECO spends close to $2 million annually to support environmental career development programs for qualified minority students.

ECO has a special program to recruit, train, and place retired engineers and scientists with nonprofit community groups reducing industrial toxics. This so-called Technical Advisor Program for Toxics Use Reduction (TAPTUR) is a companion to a service offering the talents of young persons qualified to assist communities with environmental research and consulting services.

Through Environmental Career Services, ECO supports conferences, workshops, seminars, and individual counseling—all directed towards increasing the supply of trained environmental professionals. *The New Complete Guide to Environmental Careers* is a 1993 edition of ECO's comprehensive guide, edited by Kevin Doyle and Bill Sharp. *Beyond the Green: Redefining and Diversifying the Environmental Movement* shows how people of color are playing a role in environmental management.

Environmental Career Center

"The Environmental Career Center (ECC), Hampton Virginia, is a nonprofit educational organization established in 1980 and dedicated to helping people help the environment," says Executive Director John A. Esson. "Our purpose is to lead in environmental career awareness and career development to ensure an adequate supply of trained, dedicated, and confident professionals to solve environmental challenges of the twenty-first century."

ECC's target audiences are students, advisers, recent college graduates, career changers, and persons affected by company downsizing and military base closings.

All segments of the public are reached through a twice-monthly subscription jobs bulletin, *Environmental Careers World.* The bulletin provides detailed announcements of openings at all experience levels in ecology, biology, forestry, natural resources, environmental education, science/engineering, policy and advocacy, and law—500 or more openings per month.

Colleges and universities engage ECC to conduct environmental career seminars and job fairs. A number of professional organizations, including the American Fisheries Society, Soil & Water Conservation Society, and The Wildlife Society, contract with ECC to conduct career seminars at their annual conferences.

Esson describes "the heart of our services" as career counseling, which he says "has always been free, and which we hope to continue if funding permits."

What ECC calls its Partnership Program is an arrangement with selected "partners," who have completed at least the junior year in college or technical school—though they can have graduate degrees, be midlevel professionals, even job changers or retirees. Partners are assisted in finding paid positions. Finally, in the Minority Achievement Program, interns are doing research on contamination sites in low income and minority neighborhoods. Sponsors include the U.S. Environmental Protection Agency (EPA), other civilian and military agencies, and corporations.

Educational Resources Information Center/Clearinghouse for Science, Mathematics, and Environmental Education

The Educational Resources Information Center (ERIC) system is sponsored by the U.S. Department of Education. It has developed the world's largest education-related data base, containing about 900,000 records such as articles, books, and curriculum and teaching guides. ERIC consists of 16 clearinghouses, one of which is the ERIC Clearinghouse for Science, Mathematics, and Environmental Education (CSMEE), funded by the Department of Education and located at The Ohio State University, Columbus.

ERIC/CSMEE's extensive resources are available to teachers or professors at any level, professionals, librarians/information specialists, parents, and the general public. It is said, in the literature, that students "can gain access to the latest information for preparing term papers, theses, and dissertations; obtain information on career development; and build a personalized, low-cost environmental education library."

Professor Joe E. Heimlich suggests that anyone unacquainted with the ERIC system first inquire at a college or university library or major public library for ERIC print indexes, computer access to the ERIC/CSMEE data base, or materials on microfiche. If you have a personal computer with a modem, you can access ERIC information directly or through a commercial service. Internet users can conduct individualized searches. Heimlich, who is an assistant professor of environmental education in the School of Natural Resources at Ohio State, also serves as ERIC/CSMEE's Environmental Education Associate.

Students or teachers might begin by writing ERIC/CSMEE (address at the end of this chapter) and requesting a fact sheet on CSMEE. Ask also for folders on recent publications, resources on the Internet, and articles, called digests, by Heimlich and his colleagues—for example, "Environmental Studies and Environmental Careers." These are all free and will be accompanied by price lists for other, low-cost materials.

At the technical level, ERIC/CSMEE is the distributor of documents and training manuals produced by the Environmental Protection Agency (EPA)—expressly its offices concerned with water and wastewater management.

The College Board's *Index of Majors* lists more than 350 colleges and universities offering bachelor's degrees in environmental design, environmental health engineering, or—in more than 250 cases—environmental science or studies. Many of these institutions were surveyed and asked to comment on the success of their programs and to supply literature such as catalogs. Following is a selection from their responses.

LIBERAL ARTS COLLEGES OFFERING SPECIALIZED ENVIRONMENTAL CURRICULA

Allegheny College in Meadville, Pennsylvania, 60 miles north of Pittsburgh, established an academic department of environmental studies in 1972. It now offers two interdisciplinary programs, each leading to a B.S. degree.

The major in environmental science is for students interested in fields such as environmental geology, hydrogeology, conservation biology, agroecology, pollution biology, environmental engineering, forestry, and fisheries. Required are at least 16 courses, including introduction to environmental science, a junior seminar, and a senior project. Also required are courses in biology, chemistry, geology, mathematics, and physics, with at least three upper-level electives.

The major in environmental studies is for those interested in environmental law, public policy, art, resource management, writing, and land-use planning. It includes eight core courses in environmental science, regulations, policy, and economics, and eight upper-level courses that focus on a student-selected environmental theme.

The college's promotional literature emphasizes the department's "strong groundwater component, rarely found in undergraduate studies;" its college-owned, 283-acre research reserve 7 miles from the campus; fellowship and internship opportunities; and the fact that 80 percent of recent majors were employed directly upon graduation, with 55 percent eventually attending graduate school. Allegheny also has programs (three years plus two years) providing dual degrees, a bachelor's from Allegheny and a master's from either the Duke University School of the Environment or the University of Michigan's School of Forestry. Other liberal arts colleges and major universities have similar acceleration arrangements.

College of the Atlantic is a small, private institution at Bar Harbor, Maine, where the only curriculum is Human Ecology, and 99 percent of graduates receive the degree B.A. in Human Ecology. To graduate from COA, students must earn 36 COA credits, complete a two-year interdisciplinary core curriculum, participate in a group solution to a problem, write an essay, intern, and do a major senior project.

That is the work. The pleasure is living and working on a 26-acre campus on the rocky shorefront of beautiful Mount Desert Island overlooking Frenchman Bay. Talk about small: COA began in 1972 with 4 faculty members and 16 students. It still has fewer than 250 students, two-thirds of them women, and about 25 faculty members, and does not intend to get much larger.

COA minimizes the boundaries among the disciplines, which are organized into three resource areas: environmental science, arts and design, and human studies. Each student must complete at least two courses in each of the three areas.

For environmental science, the college's setting, in close proximity to the Atlantic Ocean and bordering the Gulf of Maine and Acadia National Park, provides rich outdoor laboratories for field research. COA operates a global monitoring station, weather station, and island ecology research center, and conducts whaling and seabird studies. Greenhouses, advanced computer systems, and seagoing research vessels are available for student use.

Those preparing for graduate studies or work in plant and animal ecology, physiology, and most fields of applied environmental sciences receive both a broad-based knowledge of ecological principles, and pre-professional training in a concentration. Maine teaching certification is available, as is a Master of Philosophy in Human Ecology degree.

"Ever since its infancy, this unusual college has been brushing elbows with Harvard University," wrote Max Hall in *Harvard Magazine* (November-December 1994). "That is remarkable because it would be hard to find two institutions less alike." All four COA presidents have been Harvard graduates; that is only one example of "brushing elbows."

The Evergreen State College, a public, liberal arts college in Olympia, Washington, opened in 1971 with a mission to deliver the highest standards of teaching to undergraduate students. Over 3,000, or 92 percent, of its students are undergraduates; 12 percent are students of color. Its 1,000 acre, forested campus borders Eld Inlet on Puget Sound, and the striking new buildings have been praised by architectural critics.

The first of six features of Evergreen's "Education with a Difference" is coordinated study programs, interdisciplinary and team-taught by two- to five-member teams working with 40 to 100 students. The second feature is that a first-year student starts with a core program, also taught by a faculty team. Next is intensive exposure to a chosen specialty area, then a group contract, followed by an individual learning contract, and finally an internship.

Major curricular pathways in environmental studies include field biology and natural history; marine studies; ecological agriculture; sustainable development, political economy, and environmental policy; and geology and earth sciences.

Evergreen's literature states: "The faculty are experienced in, and committed to, providing students with practical experience through field work and projects that serve the people and organizations of southwest Washington and the Pacific Northwest."

The master of environmental studies is one of three graduate degree programs offered, along with public administration and teaching. In recognition of the Pacific Northwest's concentration of native Americans, Evergreen has a significant program of Native American Studies. Mary Hillaire, the program's principal architect, writes that the program's purpose is "to prepare learners to be able to lead a genuinely *human* life with respect to the *land, others, work* and the *unknown* in recognition of the fact that as you give, you teach others to give."

In 1966 *Prescott College* opened on a 620-acre campus a mile above sea level in the forested mountains at Prescott, Arizona. For seven or eight years, it attracted international attention as one of the new environmental colleges, but by 1974, facing bankruptcy, it closed. For two years, a few teachers and students remained, struggling against the odds, until a little money and a few more students trickled in. Prescott reestablished itself in modest, in-town quarters, and, in 1984, the regional association rewarded it accreditation at the highest level. This is believed to be the only time such academic resurrection has occurred in the United States.

Today, Prescott College is thriving as an independent four-year liberal arts college. There are about 400 resident undergraduates, and an equal number in other programs.

Environmental studies is central to Prescott's philosophy of helping people "to thrive in and enhance our world ecology." Its areas of concentration include natural history, human ecology, environmental education and interpretation, and environmental conservation. The bachelor of arts is offered, as well, in cultural and regional studies, human development, humanities, and outdoor action. Prescott offers Arizona teacher certification in all of its programs.

Beginning with wilderness orientation, Prescott students are introduced to a variety of physical, social, and cultural conditions. Advanced students lead newcomers on an expedition to an area such as the Grand Canyon for up to 19 days of backpacking. Independent studies and internships take students to Colorado for alpine ecology, Mexico for cultural ecology, and Costa Rica for tropical ecology. In 1968, Prescott College established the Center for Indian Bilingual Teacher Training to produce teacher aides with two years of college. Its purposes are to increase the proportion of certified native American teachers and to counterbalance the high failure rate of native American students.

St. Lawrence University, Canton, New York, is a private institution with 2,000 students, located in northern New York State, near the St. Lawrence Seaway, the St. Lawrence Islands National Park (Thousand Islands), and within a few miles of two Canadian provinces. Not surprisingly, the university offers a semester-long or full-year Canadian studies program. And acknowledging that rural isolation would be regarded as a disadvantage by many students, it focuses on rural environmental problems to appeal to others. Students "escape" (if that is the right word) not only to Canada but to any other of nine countries, from Austria to India, in which St. Lawrence administers international programs. Its first interdisciplinary program, established in 1973, is environmental studies.

The ES student must take a combined major, which integrates approximately seven ES courses with seven additional courses in a traditional department: anthropology, biology, chemistry, economics, geology, government, philosophy, psychology, and sociology. Core faculty numbers three, with two dozen departmental faculty offering at least one full-semester ES course.

Located in the heart of the Blue Ridge Mountains 10 miles from Asheville, North Carolina, at Swannanoa, is *Warren Wilson College,* which has a century-long heritage of service to Appalachia. This small, private college—it has only 500 students—started its environmental studies program in 1977. In recent years, ES has been the second-largest major on campus, with 60 to 70 students enrolled at all times.

This is not a school for effete city dwellers; the atmosphere is rural, and the campus includes a working farm. Each student, to graduate, must complete 60 hours of community service. Most students earn much of their room and board by working 15 hours per week on one of the more than 60 work crews devoted to campus tasks—raising hogs or broccoli, plowing the fields, building sheds or walls, or doing clerical or janitorial chores. An ES major involves disciplinary breadth, depth in a concentration selected and designed by the student, and field and laboratory experience. Most students complete two internships.

Outreach projects address local, regional, national, and international environmental needs. Twice a year students, faculty, and staff help coordinate and participate in stream cleanups. Warren Wilson teams also are conducting feasibility studies of alternate crops for Appalachian farmers. On a number of visits to a Caribbean country, work crews have built a school and a clinic.

In the mid-1990s, Warren Wilson inaugurated a wilderness-based outdoor leadership education program, and the North Carolina Outward Bound School relocated its headquarters to the campus. Graduates become wilderness instructors, operators and managers of adventure tourism enterprises, teachers, park rangers, and therapeutic recreation specialists.

U.S. UNIVERSITIES OFFERING INTERDISCIPLINARY ENVIRONMENTAL CURRICULA

Disciplinary environmental studies originated with the establishment of the land-grant colleges and universities in the mid-nineteenth century. Frequently called "agricultural and mechanical" colleges, they offered agriculture, animal husbandry, engineering, home economics, nutrition, and other applied arts and sciences. They generally came later than two other types of state institutions: teachers' colleges and state universities. The latter emphasized at their establishment, and still do, liberal arts plus such professional schools as law, medicine, and business administration.

Today, every accredited college and university, public and private, offers a broad liberal arts curriculum with basic courses applicable to an environmental career. The distinctions between "college" and "university" are blurred; for example, some of the colleges just described offer master's degrees, especially in education, often in environmental education. There follows a small sampling of universities large and small, public and private, which offer undergraduate environmental programs.

Baylor University at Waco, Texas, offers a bachelor's degree (either B.A. or B.S.) in environmental studies. What is unusual is that the degree comes as part of a double major after 28 hours of ES course work in addition to the requirements of a major in another filed. The double major provides a foundation in a traditional discipline as well as the challenge of specialized environmental study.

Many ES courses satisfy the requirements of other majors, and the introductory courses fulfill university laboratory sciences requirements for the B.A. Also offered is an ES minor consisting of 19 semester hours.

Undergraduate enrollment in the Department of Environmental Studies, established in 1969, exceeds 250, with about 40 majors and the same number of graduate students. Departmental offices, classrooms, and laboratories are located in a well-equipped building located on the banks of the Brazos River. An energy research center houses a meteorological and solar data station linked to a national network.

Santa Barbara, California, as has been noted in Chapter 1, was the site of the new environmentalism era's precipitating event: a gigantic oil spill, on January 28, 1969. If any one institution could be deemed that tragedy's direct beneficiary, it would be the *University of California, Santa Barbara.* For over a quarter of a century, UCSB has garnered international attention for excellence in environmental education and research, attracted dedicated students and faculty, been rewarded by millions of dollars of research grants, and built striking new facilities on its already breathtakingly beautiful campus overlooking the Pacific Ocean.

By the year 2000 the ES program at UCSB will have celebrated its thirtieth year as an ES entity, with over 2,300 alumni. The program is one of the strongest in student demand and national reputation. The ES program offers a bachelor of arts degree in environmental studies with a social science/humanities or a natural sciences emphasis, or a B.S. in hydrologic science.

"You can select the interdisciplinary environmental studies major, or any of a number of majors that allow you to focus on [your choice of] environmental issues," reads a UCSB brochure. "If none of the over 80 available majors suits your precise interests, you are encouraged to design your own. UCSB offers over 160 courses relating to the environment."

The UCSB environmental studies program is located in the Department of Letters and Science. The ES major prepares students for entry-level positions in many fields: urban and regional planning, environmental impact analysis, natural resource management, environmental education, journalism, conservation administration, energy policy, public interest lobbying, and government and business. It is the foundation for graduate studies in public policy; city or regional planning; architecture; social sciences; law; medicine; the physical, chemical, or biological sciences; or management. Internships often lead to first jobs or career advancement. Completion of a senior thesis is highly recommended. Specialized communications skills classes are linked to some departmental offerings each year.

Those with an interest in environmental education may wish, after completing the requirement for the B.S. in environmental studies, to pursue a California teaching credential. Advisers in the Graduate School of Education are available to assist.

Undergraduates may join a faculty research team; do an off-campus internship; and present research results at meetings or publish papers. Faculty grants and research fellowships are available to help support undergraduate and graduate students.

A graduate-level School of Environmental Science and Management is being established, with enrollment expected to reach its maximum in the academic year 1999–2000, with 75 master's students, up to 55 Ph.D. candidates, and 20 midcareer associates. A multimillion dollar building is being constructed.

Duke University, Durham, North Carolina, offers an undergraduate major in environmental sciences and policy (abbreviated EN), with courses taught by more that 60 Duke professors in 19 cooperating departments and schools. The arrangement permits students to combine interdisciplinary studies in the sciences and engineering with courses in social sciences and humanities. The major, offered since 1992, is designed for those aiming at careers in law, policy, management, or planning.

Duke has an established reputation in natural resource and environmental education dating from 1938. Now Duke has a graduate School of the Environment. It houses professional studies programs offering graduate education in environmental science, management and policy, the marine sciences, and forestry. This graduate center offers a distinctive, multidisciplinary professional and graduate curriculum focused on natural resource management, conservation ecology, environmental quality, and environmental health.

The Duke School of the Environment occupies its own new building , with state-of-the-art fiber optic networking systems linking graduate students to high-performance computing and communications resources in the nearby Research Triangle, in fact across the nation. Field studies are conducted at the Duke Forest, the Duke University Marine Laboratory, and the centers for tropical conservation, biomedicine, wetlands research, and resource and environmental policy studies.

The *State University of New York (SUNY) College of Environmental Science and Forestry* (ESF) at Syracuse is both old and new, large and small. Established as a school of forestry in 1911, it was rechartered in 1972 under the present name to reflect a broader mission in the environmental sciences and technology. With 1,800 students, ESF boasts a small-college atmosphere. Yet it is affiliated with next-door Syracuse University, which provides ESF students with the additional academic, cultural, and social benefits of a large campus. Sixty-five percent of ESF students are undergraduates.

The SUNY college has 135 faculty members, 93 percent of whom hold the highest degree awarded in a member's respective field; 68 percent hold doctoral degrees. The total book value of faculty research awards exceeds $12 million annually.

Students participate in hands-on and laboratory work at the main campus in Syracuse and on the 25,000 acres of ESF's regional campuses outside Syracuse—where the technology also is state-of-the-art. The Adirondack Forest Preserve and the Cranberry Lake Biological Field Station are sites where environmental and forest biology majors get their summer field experience.

SUNY-ESF's range of programs focusing exclusively on the environment is unmatched by any other institution in the United States. An associate-level (two-year) degree is available in forestry (resource management). The bachelor of science is available in forest engineering, paper science and engineering, wood products engineering, forestry (resources management), a dual program combining forestry and environmental and forest biology, chemistry, environmental studies, and landscape architecture—with course work, options, and elective concentrations in scores of subdivided areas.

ESF also offers preprofessional advising and study opportunities for students interested in veterinary science, medicine, dentistry, and law. In conjunction with Syracuse University, ESF offers a certifications program in secondary science teaching.

Students may enroll at many different points in their academic careers. Some may transfer in during the sophomore or junior year. For some high school students who know they want to enroll at ESF but who wish to defer enrollment until the sophomore or junior year, ESF offers a guaranteed transfer admissions plan. Formal arrangements exist with community colleges from New York State to Alabama.

Within six months of graduation, 85 percent to 100 percent of SUNY-ESF graduates in various ES/ES programs report having satisfactory employment. Over recent years, graduates have reported holding jobs— at competitive starting salaries—with 480 different job titles.

The Ohio State University School of Natural Resources (SNR) is part of the College of Agriculture, headquartered in Columbus. The school has 550 undergraduate students distributed across nine programs: environmental communication, education, and interpretation; environmental science; fisheries management and aquaculture; forestry; natural resource and environmental policy; natural resource information systems; parks, recreation, and tourism administration; soil science; and wildlife management. Also enrolled are some 80 Master of Science candidates and two dozen Ph.D. candidates.

Ohio State has a rich tradition in natural resources, dating from 1955; the current name and organization date from 1968. "Today," reports a recent bulletin, "SNR is broadening its interdisciplinary leadership within the university, expanding its international activities, and strengthening its programs to meet the needs of a changing world."

Within its modern, main building are equipment and facilities for exhibit preparation, audiovisual production, computers and laboratories for work in fisheries-aquaculture, ecosystem analysis, biochemical-nutritional studies, and wetland research. Off-campus sites include the Ohio Agricultural Research and Development Center at Wooster, the Pomerene Forest Laboratory in Coshocton, the Barnebey Center for Environmental Studies near Lancaster, and others.

A typical fact sheet defines an area; cites career opportunities; suggests salary trends; specifies high school preparation; tells how to major in the area at Ohio State; gives general education curriculum requirements and specific basic requirements for the area, a sample curriculum, and a contact point.

Although an undergraduate degree program in environmental studies per se is not offered by the *University of Wisconsin–Madison,* its Institute for Environmental Studies (IES) offers an ES certificate program. This is for students interested in environmental issues and ecological problems. Open to all UW–Madison undergraduates, the certificate program is a 26-credit, elective curriculum complementing a student's academic degree and major requirements. More than 200 students representing about 50 different majors are enrolled at any time. An exchange semester is available at the University of Guelph, Ontario, Canada.

IES offers many of its own courses and cross-lists dozens more with other academic departments. A bulletin calls instruction "the lifeblood of IES" and describes IES as "a unique intercollege unit created in 1970 to promote, develop, and administer interdisciplinary environmental instruction, research, and public service programs."

Year-round, "the institute offers more than 100 different courses... which, together attract thousands of students each year. The range of subjects is remarkable: from environmental health to environmental ethics, from natural resources to natural hazards, and from climates of the past to energy sources of the future."

While IES is not degree-granting, the University of Wisconsin–Madison offers bachelor's degrees in scores of environmental disciplines such as botany, horticulture, wildlife ecology, and zoology. Again outside IES, the Botany Department administers an interdisciplinary undergraduate degree program in biological aspects of conservation. Students in any of these programs can enroll in IES's certificate program, equivalent to a minor.

CANADIAN UNIVERSITIES OFFERING DIVERSIFIED ENVIRONMENTAL CURRICULA

The United States and Canada share not only the same continent, the same long border, and the same diversified physical environment, but also many of the same values and the same commitment to environmental education at all levels.

In the early 1990s, both countries' national legislative bodies were, at the same time, considering ways to bolster environmental education. The Environmental Ministers from Canada's ten provinces called their method the "Green Plan." Among the things called for was the integration of environmental education into all areas of learning. The goal was environmental literacy for the entire population. Canada's major universities were ready, and had been since the first Earth Day.

The School of Environmental Health at *Ryerson Polytechnic University,* Toronto, Ontario, offers the bachelor of applied arts in environmental health. Options are available in public health or in occupational health and safety, "both of which are unique in Canada." Graduates of the latter may go on to become a Registered Occupational Hygienist or a Registered Safety Professional, equivalent to the U.S.'s industrial hygienist. Although Ryerson does not offer a master's degree, it provides for university graduates a public health option leading to a second bachelor's degree. Another environmental curriculum leads to the bachelor of applied arts in urban and regional planning.

At the *University of Victoria,* Victoria, British Columbia, the environmental studies program takes three approaches: students can choose the conceptual approach, which introduces a range of environmentally related disciplines; the topical, which focuses on a particular topic; or the contract option, which allows students to develop their own programs in consultation with faculty members. The second major, for example, might be in economics, geography, or biology. Major areas of specialty are resource management, environmental planning, environmental protection, and environmental journalism.

Dating from 1969 is the program of the Department of Environment & Resource Studies at the *University of Waterloo,* Waterloo, Ontario. Recent literature outlines the program's guiding principles: It is "transdisciplinary;" undergraduate courses are organized around problem-solving, research, and systems management skills; and the "concept of learning emphasizes open-door, personal contact between students and faculty members to the extent possible."

Waterloo offers both a certificate program in environmental assessment, and an honors program (regular or co-op, in which students alternate campus studies with scheduled 4-month work terms). Among the departments with joint honors are anthropology, biology, chemistry, earth sciences, economics, English, fine arts, French, geography, mathematics, political science, philosophy, psychology, recreation, religious studies, Russian and Slavic studies, and sociology.

Waterloo has student exchange agreements with Deakin University and Royal Melbourne Institute of Technology, both in Melbourne, Australia; and with the School of Natural Resources of the University of Michigan, Ann Arbor.

WHERE SHOULD YOU WRITE FOR MORE INFORMATION?

Organizations Cited

Environmental Career Center
   (Publication: *Environmental Careers World*)
   22 Research Drive, Suite 102
   Hampton, Virginia 23666

The Environmental Careers Organization
   (Publication: *Connections*)
   286 Congress Street
   Boston, Massachusetts 02210-1009

ERIC Clearinghouse for Science, Mathematics, and Environmental Education
   1929 Kenny Road
   Columbus, Ohio 43210-1080

North American Association for Environmental Education
   (Publication: *Environmental Communicator*)
   Brukner Nature Center
   5995 Horseshoe Bend Road
   P.O. Box 400
   Troy, Ohio 45373

Colleges and Universities Cited

Allegheny College, Department of Environmental Science, Meadville, Pennsylvania 16335

Baylor University, The Department of Environmental Studies, P.O. Box 97266, Waco, Texas 76798-7266

College of the Atlantic, Registrar, 105 Eden Street, Bar Harbor, Maine 04609

University of California at Santa Barbara, Environmental Studies Program, Santa Barbara, California 93106-4160

Duke University, Environmental Sciences and Policy Program, Box 90328, Durham, North Carolina 27708-0328

The Evergreen State College, Admissions, Olympia, Washington, 98505-0002

State University of New York, College of Environmental Science and Forestry, Undergraduate Admissions, One Forestry Drive, Syracuse, New York 13210

The Ohio State University, School of Natural Resources, 210 Kottman Hall, 2021 Coffey Road, Columbus, Ohio 43210-1085

Prescott College, Resident Degree Program Admissions, 220 Grove Avenue, Prescott, Arizona 86301

Ryerson Polytechnic University, School of Environmental Health or School of Urban and Regional Planning, 350 Victoria Street, Toronto, Ontario, Canada M5B 2K3

St. Lawrence University, Environmental Studies, Canton, New York 13617

University of Victoria, Environmental Studies Program, P.O. 1700, Victoria, British Columbia, Canada V8W 2Y2

University of Waterloo, Undergraduate Adviser, Environment and Resource Studies, Waterloo, Ontario, Canada N2L 3G1

Warren Wilson College, Environmental Studies or Outdoor Leadership Studies, P.O. Box 9000, Asheville, North Carolina 28815-9000

University of Wisconsin–Madison, Student Services Coordinator, Institute for Environmental Studies, 550 North Park Street, Science Hall, Madison, Wisconsin 53706-1491

CHAPTER 3

THE SCIENCES OF LIVING ORGANISMS

We base much of what we regard as our civilization—including agriculture, forestry, and medicine—directly on our ability to manipulate the characteristics of plants, animals, and microorganisms. Thus, these discoveries have profound implications for our welfare. They teach us to utilize the productive capacity of the global ecosystem on a sustainable basis.—*Opportunities in Biology,* National Research Council.[1]

PROFESSIONS COVERED

Biology, Ecology, Environmental Chemistry, and Chemical Engineering

Many biological and biomedical scientists work in research and development. They may work in a laboratory, a library, or elsewhere—in the woods or at a desk or aboard a plane.

Some bioscientists, as they might be called, conduct basic research to increase knowledge of living organisms. Others produce better medicines, increase crop yields, and enhance environmental protection. The equipment they use may be as simple as chalk and a chalkboard or as complex as a space satellite. Some conduct experiments on laboratory

---

[1]*Opportunities in Biology,* © 1989 by the National Academy of Sciences, National Academy Press, Washington, D.C.

animals and must conform to rigorous protocols for their humane treatment. Others conduct investigations requiring human subjects, all volunteers and also under strict protocols guaranteeing the participants' rights, such as protection from harm.

Advances in basic biological and biomedical knowledge, especially at the genetic level, have spurred the field of biotechnology. Bioscientists using biotechnology techniques manipulate the genetic material of beef cattle or tomatoes, attempting to produce leaner meat and to give tomatoes a longer shelf life with more fresh flavor. Others conduct research on photosynthesis, nitrogen fixation, plant growth and development, and the ways plants interact with climate, soil, and other environmental factors.

With the expenditure of several billion dollars, the Human Genome Project is well on the way to producing a sequenced map of all the genes in the human body. Using such genetic tools, disease detectives from many medical disciplines now can identify and treat people who are in families at special risk for breast and ovarian cancer, cystic fibrosis, high blood cholesterol, and heart attacks. Recombinant DNA techniques are essential to the development of new vaccines, drugs, and cell-regulating molecules called growth factors.

Biological sewage treatment techniques, where microbes destroy disease agents, have been common for decades. Now the principle called *bioremediation* is employing microbes to "eat up" hazardous wastes at sites such as fuel spills, leaky storage tanks and pipelines, toxic soils, and contaminated groundwater. Bioremediation was directed at marine microbes in the cleanup of the 1989 *Exxon Valdez* oil spill in Alaska.

BIOLOGY

The biologist's first commitment is to the traditions of science; the second is to the profession of *biology;* and the third to the subdivision of biology in which one practices. Those are the priorities suggested by the American Institute of Biological Sciences (AIBS), which defines its field:

Biology, the most intriguing and pervasive of sciences, is the study of life and living things. It is actually a multiscience composed of many disciplines unified by the fact that all living things—plants, animals, and microorganisms—follow the same fundamental laws of heredity, reproduction, growth, development, self-maintenance, and response.

What Do Biologists Do?

Most scientists probably find their particular disciplines "the most intriguing and pervasive of all sciences" (or they would have chosen some other). Nevertheless, the thrust of the definition is factual. Most scientists are intellectually curious and honest. They enjoy posing hypotheses and testing them against observations. When satisfied with the validity of resulting data, they are eager to publish, to share the information with their peers in a scientific journal, so that others may replicate the experiments, validate (or nullify) the results, and contribute substantiating evidence and additional data. By such shared endeavors all science is advanced.

Because the area of biological sciences is so broad and so complex, it is subdivided into many categories, the major ones being outlined but not detailed here. Ecology is called a *biological science* "distinguished by its emphasis on interactions. . . and its focus on systems inclusive of life." Other specialists in the biological sciences (*biosciences,* for short) might or might not consider themselves *environmental scientists,* although often the consequences of their work have notable environmental applications and implications. The biologist who studies plants is a *botanist;* the one who studies animals is a *zoologist.* Another, who studies form and structure, including development, is a *developmental biologist.* Yet another who studies the function of whole organisms and their components is a *physiologist.* The one concerned with the heredity mechanisms which control both structure and function is a *geneticist.* One who studies the evolution and classification of plants and animals is a *systematist* or *taxonomist. Aquatic biologists* study, manipulate, or control plants and animals living in water. *Marine biologists* deal with saltwater organisms.

Biochemists study the chemical composition and processes of living things. Biochemists and *molecular biologists* are heavily involved in biotechnology.

Microbiologists investigate the growth and characteristics of microscopic organisms. Many microbiologists work in medical and public health research—in fact, some are leading in the fight against human immunodeficiency virus (HIV) and the devastating disease HIV causes, acquired immunodeficiency syndrome, or AIDS.

The American Society for Microbiology (ASM) defines its field:

> Microbiologists study microbes—bacteria, viruses, rickettsiae, mycoplasma, fungi, algae, and protozoa—some of which cause diseases, but many of which contribute to the balance of nature or are otherwise beneficial.

> Microbiological research involves recombinant DNA technology, alternative methods of energy production and waste recycling, new sources of food, new drug development, and the etiology of sexually transmitted diseases, among other areas. Microbiology is also concerned with environmental problems and industrial processes.

The ASM aggressively seeks students who are interested in its field and will furnish a free brochure, *Your Career in Microbiology: Unlocking the Secrets of Life.* By writing or sending in a response card found in the brochure, you can request additional information, the name of an ASM member to talk to, a list of colleges and universities, and other materials. (For the ASM address, see the listing at the end of this chapter.)

What Education Do Biologists Need?

To become a biologist you should have as much high school science and mathematics as possible, preferably four years of each. You should have a bachelor's degree. Virtually every college and university provides training in biology to the B.S. or B.A. level. An environmental science/environmental studies curriculum, such as those described in Chapter 2, should include much biology and may qualify you for an entry-level job. Professional jobs in biological research may be difficult for the bachelor's graduate to find. You can maximize your employability, however, by taking courses in English, chemistry, physics, and mathematics, as well as statistics and computer sciences. Training programs, such as those of The Environmental Careers Organization in Boston and the Environmental Career Center in Virginia, are especially valuable in providing practical career guidance, internships, and experience.

When you are ready and able to begin postgraduate studies, by all means do so. Graduate school need not follow immediately upon receipt of the bachelor's degree. Many students are neither psychologically ready nor financially able to go on to graduate school right after college. Take the best available job, even though it may be beneath the level for which you are qualified. It will afford useful experience, provide income to meet later graduate school expenses, and give you a "breather," a change of pace, even a chance to move to a new city or a new part of the country.

Alternate periods of work and study are often the student's best course. Such a plan lets you put theory into practice, provides practical experience essential to advancement, helps you make choices on career goals and objectives, and allows perspective on personal as well as professional decisions. Those are the reasons that cooperative education, for example, has been so successful at many schools of engineering, and why many colleges and universities require undergraduate internships.

Many employers, public and private, provide such incentives to graduate study as time off with pay, tuition and fees, and sometimes living and other expenses.

Membership in professional societies is optional, and there are no licenses or registrations required for the biologist. As in other sciences, one's peers exert pressure to publish research findings and to participate in professional activities. Even so, this peer pressure would be felt largely by the biologist on the college or university faculty, not on the one in industry or government.

Where Do Biologists Work?

Of all biologists in all specialties, somewhat over half are engaged in teaching and research at colleges and universities. The other half are divided between industry and government, and the BLS calls them "non-faculty biological scientists."

In private industry, bioscientists are employed in pharmaceuticals, chemicals, food processing, hospitals, and research laboratories.

Bioscientists work at all levels of government—federal, state, and local. In the federal government, this means the U.S. Departments of Agriculture and Interior, the armed forces, the National Institutes of Health, the National Institute of Environmental Health Sciences, the Food and Drug Administration, the Environmental Protection Agency, the Centers for Disease Control and Prevention, and literally scores of others.

Biologists are distributed fairly evenly throughout the United States and Canada because higher education and industry are so widely dispersed. But employment is concentrated in metropolitan centers, on college campuses, and in facilities such as Research Triangle Park, North Carolina.

With only a bachelor's degree, you would have limited possibilities for advancement in research and development or other professional tracks. New graduates at this level often get testing and inspecting jobs or become technical sales and service representatives. With courses in education, and teacher certification, you might become a secondary school biology teacher. A master's degree is essential for a senior position in industry or a college faculty appointment.

What Do Biologists Earn?

Supply and demand affect not only the availability of job openings but also salary and other compensation. The biological sciences constitute one of the largest areas of environmental employment: 117,000 jobs in 1992 among "nonfaculty biological scientists" tallied by the BLS, or excluding those holding faculty positions in colleges and universities.

Among the biological and life scientists considered, according to the BLS, median annual earnings for biological and life scientists in 1992 were about $34,500; the middle 50 percent earned between $26,000 and $46,800. The College Placement Council, which tracks beginning salaries in private industry, found offers that same year averaged $21,850 a year for B.S. recipients in biological science.

The BLS says in the federal government in 1993, "general biological scientists in nonsupervisory, supervisory, and managerial positions earned an average salary of $45,155...[and] ecologists averaged $44,657."

Biological scientists who have only a bachelor's degree today generally would begin in the federal government at the GS-5 level, which carries a beginning salary of approximately $18,500 per year. An entrant with a master's would probably start at Grade 7, which pays over $23,000 per annum.

Want to earn almost twice that amount, in Alaska, with "strenuous working conditions, but supportive company"? The company advertising is a consulting firm, which offers a training class every month, up to $3,500 per month, and year-round opportunities. Duties would be observing, collecting, sorting, and recording sample catches on board commercial fishing boats. Qualifications include a B.S. in biology "and a flexible attitude"—no experience required.

The Nature Conservancy, a major conservation group, often seeks conservation workers at various locations—Sioux Falls, South Dakota; Honolulu, Hawaii; Charlottesville, Virginia; and others. Some, but not many, of these openings are for beginners.

A new field for the sales-and-promotion minded individual is that of "environmental activist," a common euphemism for fundraiser or solicitor. Greenpeace, Clean Water Action Project, Sierra Club, and Public Citizen's Critical Mass Energy Project are among the national groups that raise significant portions of their budgets by door-to-door or telephone solicitation. Jobs are to be found through classified ads in many metropolitan newspapers. Students may do this during summer vacations and may join teams that travel from city to city for special campaigns. Offerings start at about $16,000 and seldom go above $24,000. Remuneration is generally based on commission, that is, a percentage of funds raised.

In Canada, salaries appear to approximate U.S. levels. However, figures provided by three Canadian national agencies all seemed to be for experienced professionals rather than entry-level personnel. The Canadian Forest Service, for example, reported that, in 1995, senior scientists/biologists had an average salary of $53,414 in Canadian dollars.

(Relative to the U.S. dollar, the Canadian dollar fluctuates between eighty cents and eighty-five cents; i.e., a given salary in Canada is 15 percent to 20 percent less than the equivalent salary in U.S. dollars.)

Program managers and other midlevel personnel in Canada would have an average annual salary of $49,352. More junior would-be wardens and interpreters average $41,719. New graduates of universities or technical schools might be paid at an hourly wage, which would "vary widely," said a Canadian official, but which typically would be about $34,091.

This scale seems markedly higher than the U.S. Labor Department's Bureau of Labor Statistics's reports for U.S. salaries. However, it should be taken as a rough translation of hourly wages without benefits—an area in which the U.S. government and U.S. corporations may be more generous than their Canadian counterparts.

What Is the Job Future for Biologists?

Employment of biological scientists is expected to increase faster than the average for all occupations through the year 2005, according to the BLS, which for the first time projects employment trends into the twenty-first century.

The AIDS epidemic and new methods of gene therapy are among the factors driving the demand for more microbiologists for health research. The National Institutes of Health (NIH) at Bethesda, Maryland, is advertising for microbiologists in all these areas. However, new B.S. graduates should not apply for these positions; they almost always require experience. But NIH has several programs for summer employment of high school and college students.

Salaries for experienced M.S. or Ph.D. chemists, biologists, microbiologists, and others are on the government-wide (GS) scale, with a Washington-area locality add-on now amounting to 5.48 percent. At the GS-13 level, these jobs pay up to $67,000 per annum. Also advertised are microbiologist positions at the GS-7, 9, and 11 levels, paying (in 1995) $24,000 to $36,000 per annum.

Several times a year, *Science* magazine features special editorial sections on career opportunities for various categories of scientists:

women, minorities, physically handicapped, and those with the B.S. or M.S. as their highest degree. These jobs invariably require experience, but the student considering a career in the topic area covered may find both the editorial and the advertising content helpful and interesting in career planning. The weekly *Science* is available in any college library and in many public libraries.

The BLS's crystal ball shows:

> Biological and biomedical scientists will continue to conduct genetic and biotechnological research and help to develop and produce products developed by new biological methods. In addition, efforts to clean up and preserve the environment will continue to add to growth. More biological scientists will be needed to determine the environmental impact of industry and government actions and to correct past environmental problems.

Here, however, the BLS has a significant caution: "[M]uch research and development is funded by the federal government. Anticipated budget tightening should lead to slower employment growth of biological and medical scientists in the public sector and in some private industry research laboratories as the number and amount of government grants and contracts increase more slowly than in the past."

Fortunately, biological scientists are flexible. Summer jobs, personal and family contacts, and teacher recommendations count heavily in gaining entry and experience. Many persons with a two-year associate's degree or a four-year bachelor's degree can find jobs as science or engineering technicians or health technologists in a community hospital. Some become high school biology teachers.

Once established, the bioscientist enjoys job mobility. Those with doctorates or M.S.'s often can move from government to academe, from industry to government, from agricultural research to biotechnology, from country to city, or vice versa. It is becoming more common in all fields to switch employers, and to retrain for career change, even for different career paths.

Biological scientists may be less likely than others to lose their jobs during recessions. If they teach at the college level, they should have tenure. If they do research in government, they have a significant amount of security. If they are dependent upon research contracts or grants and these are cut back, the assistants, technicians, and graduate students will feel the blow much earlier and much greater than the associate professor or professor. Of course, a recession reducing the amount of money allocated to new research and development efforts would affect everyone in the system.

On a more long-range basis, *Opportunities in Biology* predicts a bright future for the field. One area is *biotechnology,* defined by the National Science Foundation and the Office of Technology Assessment as a technique that uses living organisms or parts of organisms to make or modify products, to improve plants or animals, or to develop microorganisms for specific uses.

The five main areas of research and development in biotechnology are health care, plant agriculture, chemicals and food additives, animal agriculture, and energy and the environment.

An exciting early application of biotechnology is in biomedicine. The human gene that codes for the production of insulin has been inserted into bacteria, causing the bacteria to produce human insulin. This insulin, used to treat diabetes, is much purer than insulin from animals, the only previous source. Other substances not previously available in large quantities are starting to be produced by biotechnological means; a number are on the brink of approval for treating cancer and other diseases.

Of special interest to environmentalists is another new area, *conservation biology,* whose responsibility is understanding, and preventing the crises in, habitat degradation and species extinction.

About conservation biology and those who practice it, *The Scientist* recently wrote:

> Although these researchers and their colleagues work in far-flung locales and on a menagerie of species, they have one thing in common: Their investigations are part of a relatively new marriage of fields—molecular and conservation biology. By applying such techniques as the polymerase chain reaction (PCR) and DNA fingerprinting, they use some of the smallest biomolecules—nuclear and mitochondrial DNA—to attack very large problems.

Their research efforts are targeted at maintaining the biodiversity and understanding the life history and evolution of endangered species.

If conservation biology is one of biology's newest subspecialties, microbiology is one of its oldest—and proudest. Microbiology boasts some of the most illustrious names in the annals of science—among them Pasteur, Koch, Fleming, Lister, Leeuwenhoek, Jenner, and Salk—and some of the greatest achievements of all time. During the twentieth century, one-third of all Nobel Prizes in Physiology or Medicine have been bestowed upon microbiologists!

Addressing young persons, the American Society for Microbiology offers this challenge:

> As Dr. Ronald Cape, chairman and founder of a large biotechnology company, says, "There are lots of Nobel prizes waiting to be earned!"
>
> Your choices are not only exciting, but vast. Projections for the next 20 years indicate thousands of scientific positions will go unfilled with a huge demand for microbiologists. A variety of special programs will make opportunities [especially favorable] for women, members of underrepresented groups, and people with disabilities.
>
> Where you end up is your choice. Your future is bright and promising. With a solid foundation in science and the desire to discover an unknown world, your possibilities and opportunities in microbiology are numerous.

(Aside from the Nobel, there are several environmental science awards of great distinction. The Tyler Prize for Environmental Achievement carries an honorarium of $150,000, and The Heinz Award for Environmental Contributions includes a cash prize of $250,000.)

ECOLOGY

A group of living components in a natural neighborhood constitutes an ecological system, an *ecosystem,* such as a forest, a lake or an estuary. Larger ecosystems—which occur in a similar climate and have similar vegetation—are called *biomes;* examples are the arctic tundra, prairie grassland, or desert. The *biosphere* is comprised of the earth, with its surrounding envelope of life-sustaining air and water, plus all its living things. All of these are elements of the *environment,* including as well people and their social, political, and economic systems. *Ecology* is the discipline concerned with all such features and their connections.

The Ecological Society of America provides this definition:

> Ecology is the study of interactions among all forms of life, and between organisms and their environments. Ecology is distinguished from other biological sciences by its emphasis on interactions, and from other environmental sciences by its focus on systems inclusive of life.

What Do Ecologists Do?

The ESA further says:

> Ecologists may focus on the natural history of a species of fish or insect, use the data to develop a mathematical theory of patterns of species distribution, or may seek to understand the relationship between species diversity and magnitude of pollutants in rivers and lakes. Whatever the focus of study, ecologists seek to understand the basic processes which have formed and maintained these systems and have created their special character.

The majority of ecologists, being on university faculties, teach and conduct research. As administrators of their research projects, they are responsible for keeping books and records which are audited as required by laws and regulations.

Ecologists frequently work in teams; their activities depend on cooperation with and support from others, and the team's results depend greatly on the integrity of each ecologist's contribution.

What Education Do Ecologists Need?

The Ecological Society of America, in its careers folder, stresses the importance of early preparation in high school. Take a well-rounded program including biology, mathematics, physics, geology, chemistry, social sciences, and humanities, ESA advises. In college, you might major in the biological sciences, taking courses in morphology, physiology, and genetics, as well as ecology. Inorganic and organic chemistry are essential, and biochemistry and physical chemistry may be important for certain areas of ecology. If you aspire to a research career—and many ecologists do—you should take as much physics and mathematics as possible, including calculus, linear algebra, probability theory, statistics, computer science, and economics. French and German are recommended as the languages of choice, with Russian and Chinese growing in popularity.

In your high school or college years, you should be able to find summer jobs related to ecology, and thereby judge whether you have sufficient interest and aptitude to do well in this field. For summer jobs, first check your career office's bulletin board, and then your local newspaper. The *Newsletter of the Ecological Society of America* regularly announces undergraduate as well as postgraduate research assistantships and volunteer opportunities nationally (and is generous in making the newsletter widely available). A recent issue advertised two summer assistantships in zooplankton ecology at Southwest Missouri State University at Springfield, and three assistantships in conservation and sustainable development for summer work in Costa Rica for Cornell University. Volunteers are wanted by The Nature Conservancy, New Mexico Chapter in Santa Fe. (For a free subscription to ESA's newsletter, see the listing at the end of this chapter.)

Most positions for ecologists advertised in that newsletter, in *Science* magazine, or *ES&T: Environmental Science & Technology* require the M.S. degree if not the Ph.D.

Specialization can begin during undergraduate years, with a major in biology, botany, or zoology, with special preparation in general ecology. But a B.S. degree is not sufficient; you should expect to go on for a master's and eventually a doctor's degree. Most professional ecologists hold the Ph.D.

The Ecological Society of America grants professional certification to those who qualify at each of three levels, requiring a successively higher educational degree or depth of experience. The levels are called associate ecologist, ecologist, and senior ecologist.

Where Would You Work as an Ecologist?

Teacher, researcher, or administrator—the ecologist most often fills all three roles simultaneously. Some do research exclusively, as employees of the federal or state government, frequently at an agricultural experiment station of a land-grant college. Still others work for private companies in forestry, paper products, or large agricultural operations. Moreover, ecologists study the impacts of energy developments on the environment. Oil and natural gas producers have ecologists on their exploration teams, along with geologists and other earth scientists. Such teams are studying the outer continental shelf, coastal zone, estuaries, and even desert areas to establish baseline conditions before new facilities disrupt the ecology, to monitor the environment while new facilities are being built, and to assure that problems do not arise after these facilities are in operation. Any industry or utility that is planning to build, for example, a power plant in a fragile environment needs to employ an ecologist for such studies. Planning, zoning, and regulating bodies require environmental impact studies (EIS), which may be provided, or contributed to, by ecologists. These contributors may be on the payroll of an engineering or biological consulting firm, or they may be graduate students or professors.

In a recent issue of ESA's newsletter, EG&E Energy Measurements, Inc., Las Vegas, Nevada, advertised for a candidate with a bachelor's degree (M.S. preferred) "to coordinate field surveys for the threatened desert tortoise and candidate species for federal listing under the Endangered Species Act. Additional responsibilities include directing field personnel, preparing schedule and cost estimates, writing technical reports, and maintaining data records and computer data bases." The salary range was $32,000 to $50,000 per year. That is typical of the ads this particular government contractor, which needs many people for hot, dusty work, has been running for five years or more.

Most basic research in ecology is sponsored by the National Science Foundation through grants to universities. Other government agencies—primarily the Departments of Agriculture, Interior, and Defense; the Environmental Protection Agency; and the National Oceanic and Atmospheric Administration—are major employing agencies. The United Nations Environment Program, various specialized U.N. agencies, and private philanthropic foundations employ ecologists in international programs.

What Do Ecologists Earn?

Ecology is a small field, the smallest component of the biological sciences. More than half, perhaps a majority, of all ecologists work directly for the federal government. They got their jobs through competitive examination, which means an evaluation of their records compared to other applicants applying for similar jobs at the same time in the same geographical area.

With a bachelor's degree, one would usually begin in the government at the General Schedule grade 5 (GS-5), with a beginning annual salary in the mid-1990s of approximately $18,500 per year. An entrant with the master's would probably start at grade 7, which pays over $23,000 a year. Academic standing and research accomplishments—as well as recommendations of professors, employers, or supervisors—could qualify one for a single-grade advance, amounting to perhaps two thousand dollars more. Within each grade there are ten levels paying progressively higher salaries for satisfactory performance. The latest annual increase for federal civilian employees, recommended by the president and approved by Congress, was 3.22 percent, which is about average over recent years.

What Is the Job Future for Ecologists?

The Bureau of Labor Statistics (BLS) lumps ecology into the biological sciences; therefore, BLS provides no separate report on employment, or employment prospects, for ecologists alone. As ecology is a component of the biological sciences, it grows and thrives as its larger cousin, biology, grows and thrives.

However, it is estimated that there are approximately 8,000 ecologists in the United States, and turnover, replacement, and expansion create some 400 new openings per year.

More than likely, an environmental science/environmental studies (ES/ES) bachelor's degree from any liberal arts college (some programs more than others) should qualify you for an entry-level job in ecology. And it is certain that an ES/ES bachelor's from any major university would qualify you.

The *Newsletter of the Ecological Society of America* suggests a continuing if only a slowly growing demand for ecologists with the B.S. degree alone, greater demand for M.S. ecologists, and a significantly greater demand for those with Ph.D.s.

ENVIRONMENTAL CHEMISTRY AND CHEMICAL ENGINEERING

Chemistry is the science of the composition, structure, properties, and reactions of matter, including its atomic and molecular systems. Chemistry often is considered an environmental science in its own right, and the American Chemical Society publishes the journal *ES&T: Environmental Science & Technology*. An environmental emphasis may enter into any of the major subfields of chemistry, as well.

Environmental chemical engineering is a combination of chemistry and engineering, whose practitioners take an engineering approach to the technical problems of their employers.

What Do Environmental Chemists and Chemical Engineers Do?

The *analytical chemist* determines the structure, composition, and nature of substances and develops new techniques. Originally, the *organic chemist* studied the chemistry of living things, but this area has been broadened to include all carbon compounds. When combined with other elements, carbon forms an incredible variety of substances. Many modern commercial products, including plastics and other synthetics, have resulted from work in organic chemistry. The *inorganic chemist* studies compounds other than carbon and may develop, for example, materials for use in solid-state electronic components. The *physical chemist* studies energy transformations to find new and better energy sources. The *toxicologist* conducts tests on animals to determine the effects of drugs, gases, poisons, pesticides, radiation, and other substances on the health of the organism. The *biochemist* or *biophysicist* studies the chemical and physical behavior of living things. Because life is based on complex chemical combinations and reactions, the work of the biochemist is vital for an understanding of the basic functions of living things, such as reproduction and growth. The biochemist also may investigate the effects of substances such as food, hormones, or drugs on various organisms.

What Education Do Environmental Chemists or Chemical Engineers Need?

Nearly every college and university offers a basic degree in chemistry. Only selected engineering colleges offer a chemical engineering degree. More than 350 institutions offer graduate degrees in chemistry. Graduate students generally are required to have a bachelor's degree in chemistry, biology, or biochemistry. Many graduate schools emphasize some specialties of chemistry over others because of the type of research being done at those institutions. If you lean toward a certain type of environmental career, pick your graduate school with that leaning in mind. Graduate training requires actual research in addition to advanced science courses. For the doctoral degree, you specialize in one field of chemistry by doing intensive research and by writing a dissertation.

Many schools or colleges of engineering include chemical engineering among the subspecialties offered. A bachelor's in chemical engineering is the standard professional degree, and the master's is becoming more common. The doctorate is required only for those who aspire to become engineering-school professors.

Where Do Environmental Chemists or Chemical Engineers Work?

Chemists held about 92,000 jobs in 1992, according to the BLS, and chemical engineers about 52,000. The majority of chemists, and 72 percent of the chemical engineers, worked in the chemical manufacturing industry or in food processing. Chemists also work for state and local governments, primarily in health and agriculture, and for federal agencies, chiefly the Departments of Defense, Health and Human Services, and Agriculture. About 20,000 chemists are on the faculties of colleges and universities.

Jobs are distributed throughout the United States and Canada, especially in metropolitan areas, on college campuses, and at agricultural and other research stations.

Chemical engineers work primarily in the chemical and petroleum-refining industries, but also for engineering firms or as independent consultants. For them, opportunities abound to travel and live overseas, especially in the oil industry in oil-producing regions such as the Persian Gulf.

How many of the total population would be defined as *environmental* chemists/chemical engineers is problematical. The BLS does not use the modifier *environmental,* so any estimate would be guesswork, but both the American Chemical Society—whose members are from both professions—and the American Institute of Chemical Engineers consider it substantial.

What Do Environmental Chemists and Chemical Engineers Earn?

According to a 1992 survey by the American Chemical Society, the median starting salary for recently graduated chemists with a bachelor's degree was about $24,000 per year; with a master's degree, $50,000; and with a Ph.D., $60,000. In 1993, the BLS reports, chemists in the federal government earned an average salary of almost $52,000.

Starting salaries for engineers with the bachelor's degree are significantly higher than for college graduates in most other fields. According to the College Placement Council, chemical engineers with the bachelor's degree, winning first jobs in industry, had an average starting salary in 1992 of $39,203. This was the second highest of ten types of engineers, exceeded (by $1,500) by petroleum engineers, and almost $10,000 greater than the average starting salary for the lowest-paid, civil engineers.

Chemists and chemical engineers frequently advance to management positions. Here, the salary levels are significantly higher than—sometimes more than twice as high as—any cited above. Major chemical industry corporations employ environmental department directors, often at the vice-presidential level, and at very high salaries commensurate with their enormous responsibilities.

Increasingly, many large chemical companies are offering a dual-track career plan. One track is for management and marketing positions; this is the traditional career ladder to higher-paying positions. What is new is the so-called technical ladder. Qualified persons can choose to remain working at the laboratory bench, doing research, publishing, competing only with their peers for promotions, getting promoted regularly to higher salaries, enjoying opportunities to travel, and participating as officers in national or international scientific societies.

What Is the Job Future for Environmental Chemists and Chemical Engineers?

An environmental science/environmental studies degree from a liberal arts college probably would not qualify you for even an entry-level job in chemistry, and definitely not for a chemical engineering, or any other professional, engineering job. Only a major in chemistry or chemical engineering would provide the credential.

"Employment of chemists is expected to grow about as fast as the average for all occupations through the year 2005," we find (again) in the *Occupational Outlook Handbook*. "However, employment will not grow as rapidly as in the past because, overall, research and development budgets are expected to grow more slowly compared to the 1980s."

The BLS is a little more optimistic about the prospects for chemical engineers:

Chemical engineers should find favorable job opportunities. The number of positions arising from employment growth, which is expected to be as fast as the average for all occupations through the year 2005—and the need to replace those who leave the occupation—should be sufficient to absorb the number of graduates [who come out of school] with degrees in chemical engineering.

The BLS makes a further, pertinent point about a large industry such as chemicals: Even when it is growing far more slowly than a small field such as ecology, the smaller percentage of openings created amounts to a large total number.

A recent article in *Chemecology,* a Chemical Manufacturers Association publication, asks, "What can you do with a chemistry degree?" It quotes Professor T.C. Ichniowski, of Illinois State University, who says that an undergraduate or advanced degree in chemistry "can be a ticket to a career in business, sales, journalism, or law." He mentions having read that a Chicago patent law firm was seeking bachelor-level chemists to provide their lawyers with guidance in biotechnology and bioengineering. And he especially recommends that chemists consider teaching in elementary and secondary schools, if they are willing to take teachers' salaries.

WHERE SHOULD YOU WRITE FOR MORE INFORMATION?

American Association for the Advancement of Science
(Publication: *Science)*
1333 H Street, N.W.
Washington, D.C. 20005

American Chemical Society
(Publication: *ES&T: Environmental Science & Technology)*
1155 16th Street, N.W.
Washington, D.C. 20036

American Institute of Biological Sciences
(Publication: *Bioscience)*
Office of Career Service
730 11th Street, N.W.
Washington, D.C. 20001-4521

American Institute of Chemical Engineers
(Publication: *Chemical Engineering)*
345 East 47th Street
New York, N.Y. 10017

American Society for Microbiology
Board of Education and Training
1325 Massachusetts Avenue, N.W.
Washington, D.C. 20005

Botanical Society of America
    The Manager of Publications
    c/o Department of Botany
    The Ohio State University
    1735 Weil Avenue
    Columbus, Ohio 43210

Chemical Manufacturers Association
    (Publication: *ChemEcology)*
    2501 M Street, N.W.
    Washington, D.C. 20037

Ecological Society of America
    (Publication: *Newsletter of the Ecological Society of America)*
    2010 Massachusetts Avenue, N.W.
    Washington, D.C. 20036

SUBSCRIPTION NEWSLETTERS LISTING CURRENT OPENINGS IN ALL ENVIRONMENTAL FIELDS

Environmental Careers World
    Environmental Career Center
    22 Research Drive, Suite 102
    Hampton, Va. 23666

Environmental Opportunities
    P.O. Box 788
    Walpole, N.H. 03608

The Job Seeker
    Route 2, Box 16
    Warrens, Wisc. 54666

GUIDES TO A JOB SEARCH

(Available at libraries)

Conservation Directory, published annually by National Wildlife Federation,
    1412 16th Street, N.W., Washington, D.C. 20036.
Occupational Outlook Handbook, published biennially by Bureau of Labor
    Statistics, U.S. Department of Labor, and *Occupational Outlook Quarterly,*

both for sale from Superintendent of Documents, U.S. Government Printing Office, Washington, D.C. 20402-9325.

Occupational Outlook Handbook reprints are available at low cost as bulletins in the 2450 series; for ordering information or to place orders, write Superintendent of Documents, U.S. Government Printing Office, at the address above, or P.O. Box 371954, Pittsburgh, Pa. 15250-7954.

CHAPTER 4

ENVIRONMENTAL HEALTH

The most difficult challenges for environmental health today come not from what *is known* about the harmful effects of microbial agents; rather they come from what *is not known* about the toxic and ecologic effects of the use of fossil fuels and synthetic chemicals [and other polluting agents] in modern society—*Healthy People 2000,* 1990.

PROFESSIONS COVERED

Environmental Engineering, Environmental Health Sciences, Environmental Medicine, Public Health Veterinary Medicine, Environmental Sanitation, Industrial Hygiene

Environmental health is defined as a rational effort that prevents epidemics and the spread of disease, protects against environmental hazards, prevents injuries, promotes and encourages healthy behaviors, responds to disasters and assists communities in recovery, and assures the quality and accessibility of health services.

The definition was developed by the Essential Public Health Services Work Group of the Core Public Health Functions Steering Committee, coordinated by the Office of Disease Prevention and Health Promotion, Department of Health and Human Services, in Fall 1994. It conforms to the task force's vision of "healthy people in healthy communities" and the modern mission of public health services: "to promote physical and mental health and to prevent disease, injury, and disability."

In addition to the statement's five objectives, it lists many essential public health services; environmental health is integral to all of them.

"HEALTHY PEOPLE 2000"

The "vision and mission statement" about public health in America is confined to one page—a rare example of bureaucratic brevity. It is based on a document which is anything but brief: *Healthy People 2000* is a 700-page report subtitled "National Health Promotion and Disease Prevention Objectives," which was issued by the Secretary of Health and Human Services in 1990. This was an update of a report entitled *Healthy People,* which was issued a decade earlier and had been regarded as very important. The staff and consultants working in 1990 had the approaching turn of the century as a goal by which to measure progress against certain diseases and health conditions generally. The report details 300 specific, measurable health objectives described as achievable by the year 2000, hence the title of the report.

Environmental health, with 16 specific objectives, forms a major component of this program. All objectives are aimed at numerical targets to be reached by the year 2000, measured from baselines between 1987 and 1991. They include reducing morbidity from asthma, mental retardation among school-aged children, outbreaks of waterborne disease, the prevalence of high blood lead levels in small children, and human exposure to air pollutants and toxic agents.

Water pollution and solid waste–related water, air, and soil contamination also would be reduced. Drinking water supplies would be improved. Homes would be tested for lead-based paint contamination and radon concentrations. Programs for recyclable materials and household hazardous waste would be extended to counties now lacking them.

Three years into the decade of the nineties, 6 percent of all 300 national health objectives for the year 2000 already had been achieved, seven years early (*Healthy People 2000 Review 1993,* 170 pp., Hyattsville, Md.: Centers for Disease Control and Prevention, National Center for Health Statistics, 1994). Progress also had been made on another 36 percent. Overall, the greatest progress was being made against three major health problems: cancer (due to healthier life-styles such as smoking avoidance, more exercise, and better nutrition, plus earlier cancer detection and new methods of diagnosis and treatment); overall disease surveillance and data systems (the traditional public health system at work); and the reduction of unintentional injuries or accidents (due to safer cars, driver education, and reduction of drunk driving, among other factors).

The record for environmental health achievements does not compare. Eight, or half, of the national environmental health objectives showed some progress in three years, but four of the other objectives "showed movement in the wrong direction." Three of the four "wrongs" are concerned with water: Waterborne disease outbreaks are worse than at the differing baselines, the proportion of people receiving safe drinking water declined slightly, and the proportion of "impaired" surface water increased. The number of hazardous waste sites on the national priorities list also grew.

In such inadequacies lies an enormous challenge to all professionals in environmental health.

PROFESSIONALISM

There are several marks that are especially important in environmental health whereby one becomes, and is identified as, a professional. The most common is membership in a professional association, a privilege usually accorded anyone who graduates from an accredited college or university with a degree in the discipline. The principal professional organizations listed at the end of this and other chapters are examples. These associations provide services such as professional education and information to their members, career information to students, and general information and services to the public. They represent the profession in dealing with other professional organizations, with regulatory and licensing bodies, and in lobbying in Washington and state capitals.

Certification is the process by which a nongovernmental agency or association (such as a professional organization just mentioned)

grants recognition to an individual who has met certain predetermined qualifications.

Registration is the process by which qualified individuals are listed on an official roster maintained by a governmental or nongovernmental agency. Within some professions there are specialty boards or registries established by the professions themselves. These boards identify those members of the profession who meet certain requirements of education, experience, and competence, as determined through examination. Such individuals may be accorded the title of "Fellow" or "Diplomate."

Licensure is the process by which an agency of government grants permission to persons meeting predetermined qualifications to engage in a given occupation which directly affects the health and welfare of the public. Sometimes the license permits these persons to use a particular title. More than 30 occupations in the health field are licensed in one or more states. All fifty states and the District of Columbia require licensure of civil engineers, environmental health engineers, and others when they are engaged in engineering work which may affect life, health, or property—and who offer their services to the public on a fee basis. The initials P.E. after a person's name stand for Professional Engineer, a mark of professional distinction. All states, territories, and the District of Columbia also require licensure for physicians (Medical Doctor, M.D., and Doctor of Osteopathy, D.O.).

ENVIRONMENTAL ENGINEERING

As an *environmental engineer,* you would command the specialized engineering knowledge essential to solve environmental problems which involve engineering. Because so many environmental problems are *interdisciplinary/integrated/involved,* increasingly the engineer is the key member of the management team.

Engineering is defined as the means by which the properties of matter and the sources of energy in nature are applied to practical purposes. The first engineers in America held a combination of civil and military roles. For example, George Washington is regarded as the father of the engineering profession as well as the father of his country. Civil engineers laid out plans for cities and designed and supervised the construction of roads, harbors, tunnels, and bridges. Then, as now, civil engineers with additional specialized training in sanitation, *sanitary engineers,* were responsible for designing and building safe public water supplies and wastewater treatment facilities and controlling flying/ crawling vectors of disease such as flies and mosquitoes. By controlling the mosquitoes which carry malaria and yellow fever, sanitary engineers made possible the building of the Panama Canal where the French had failed because of those diseases' toll. Sanitary engineers designed the sanitation facilities which have made our cities the world's healthiest urban environments.

Today, most engineers specialize; professional bodies recognize more than 25 specialties. Within the major branches are numerous subdivisions. While many practitioners of environmental engineering are registered *civil engineers*—environmental engineering being a subset of civil—others are qualified as, and go by the title of, *environmental engineers.*

What Do Environmental Engineers Do?

In its latest recruiting brochure "Our Past, The Present, Your Future...in Civil Engineering," the American Society of Civil Engineers (ASCE) highlights some of the duties:

> Environmental engineers translate physical, chemical, and biological processes into systems to destroy toxic substances, remove pollutants from water, reduce nonhazardous solid waste volumes, eliminate contaminates from the air, and develop groundwater supplies. In this field, you could be called upon to resolve problems of providing safe drinking water, cleaning up sites contaminated with hazardous materials, disposing of wastewater, and managing solid wastes.

Several "green" projects are portrayed in the colorful brochure. One is the $6 billion cleanup effort in Boston Harbor centered on the Deer Island wastewater treatment plant. Another, in the Mojave Desert, is a new solar energy generating system incorporating computer-controlled mirrors, designed to supply enough power to meet the needs of over one million Southern California residents.

Since 1988, the American Academy of Environmental Engineers (AAEE) has sponsored an annual competition to identify and recognize outstanding "projects which exemplify quality in all facets of environmental engineering practice." Cited in a recent year were a solid-waste management facility in Phoenix, Arizona; a pipeline in Norfolk, Virginia; remediation of hazardous waste materials at Rocky Mountain Arsenal, Colorado; a Superfund site remediation at King of Prussia, Pennsylvania; and clean fuels from landfill gas at Los Angeles County, California.

ASCE emphasizes that civil engineers are among the leading users of today's sophisticated high technology. The latest concepts in computer-aided design (CAD) are employed during design, construction, project scheduling, and cost control phases of any complicated construction project.

What Education Do Environmental Engineers Need?

According to the Bureau of Labor Statistics, about 390 colleges and universities offer a bachelor's degree in engineering, and nearly 300 institutions offer a bachelor's in engineering technology. Either degree normally takes four years, but many students find that there is so much to study that it frequently takes one to two years longer to complete their baccalaureate degrees.

In a typical engineering curriculum, the first two years are spent studying basic sciences (mathematics, physics, and chemistry), introductory engineering, and the humanities, social sciences, and English. In the final two or three years, most courses are in engineering, usually with a concentration in one branch.

John M. Buterbaugh of the staff of the American Academy of Environmental Engineers (AAEE), reports: "Currently, there are 95 colleges and universities in the United States that offer graduate programs in environmental engineering, and 15 with undergraduate programs for the specialty. Most baccalaureate programs in civil and chemical engineering allow students to concentrate their elective courses on environmental engineering if they wish."

All 50 states, the District of Columbia, and the U.S. territories require registration for engineers whose work may affect life, health, or property, or who offer their services to the public. In 1992, there were nearly 380,000 registered engineers, of which civil engineers numbered about 173,000. The Accreditation Board for Engineering and Technology (ABET) is primarily responsible for monitoring, evaluating, and certifying the quality of engineering education in colleges and universities in the United States. Registration generally requires a degree from an engineering program accredited by ABET, four years of relevant work experience, and passing a state examination.

The American Academy of Environmental Engineers (AAEE), founded in 1955, pioneered specialty certification for the engineering profession patterned on regularly accepted professional practices of the medical community. It is a certification and membership organization whose sponsoring organizations include the American Institute of Chemical Engineers, American Public Health Association, American Society of Civil Engineers, National Society of Professional Engineers, and a number of others.

The AAEE provides certification that the engineers who meet its requirements for education, licensure, and experience are peer-recognized specialists in one or more of seven environmental practice specialties: air pollution, general environmental engineering, hazardous waste, industrial hygiene, radiation, solid waste, and water supply/wastewater engineering.

The minimum prerequisites for certification by AAEE include a baccalaureate degree in engineering or a related field, a currently valid engineering license issued by a state, and eight years of progressively responsible engineering experience. (Some, but not all, of these criteria distinguish this certification from that of the civil engineer or other professionals.)

Candidates for certification by AAEE must pass written and oral examinations. To apply for certification, the candidate must be designated a "P.E.," professional engineer. Following certification, the designee also may use the initials "DEE," signifying Diplomate, Environmental Engineering. AAEE has more than 2,500 board-certified environmental engineers on its rolls, which are growing steadily.

Where Do Environmental Engineers Work?

William C. Anderson, P.E., DEE, provided a special section on environmental engineering employers in a 1994 issue of *Engineering News Record.* The American Academy of Environmental Engineers sponsored it, along with the Hazardous Waste Action Coalition and the Environmental Business Council of the United States, Inc. Anderson suggested that those who aspire to a career in environmental engineering might look to such firms as the section's corporate advertisers. They included engineering firms headquartered in Buffalo, New York; West Chester, Pennsylvania; Cambridge, Massachusetts; Irvine, California; Harrisburg, Pennsylvania; Dallas, Texas; Atlanta, Georgia; and every other section of the country.

According to the Bureau of Labor Statistics, over 40 percent of all civil engineering jobs are in federal, state, and local government agencies. Over one-third are in consulting engineering firms. The construction industry, public utilities, transportation, and manufacturing industries account for most of the rest.

The traditional employers of environmental engineers, according to the Association of Environmental Engineering Professors, have been local governmental units such as city or county health departments, state governments, and federal agencies. Federal laws such as the Toxic Substance Control Act, Resource Conservation and Recovery Act, and the Comprehensive Environmental Response, Compensation, and Liability Act—better known as Superfund—have created job opportunities both in federal agencies and in the private sector. The greatest job growth recently has been in consulting engineering firms and in the environmental engineering departments of industrial firms. They employ environmental engineers for research and studies on contract, to prepare and evaluate environmental impact statements, and to assure compliance of their client companies with the myriad governmental laws and regulations to which these firms are subject.

Interested in adventure overseas? The armed services offer some of the more attractive opportunities for environmental engineering practice. Increasingly, consulting engineering and construction firms with contracts with the oil-producing nations of the Middle East advertise for environmental engineers, among others. They are building, in addition to oil and natural gas facilities, whole new cities, industrial complexes, ports, university campuses, highways, and airports—in fact, entire new "developed" nations are springing up on the desert. Consequently they need significant numbers of environmental engineers.

The developing nations of Africa, South America, and the Pacific also are undertaking enormous development projects, funded by international lending agencies such as the World Bank or the Asian Bank. The work actually is done by multinational engineering and construction companies—U.S., British, Dutch, Japanese, or others. One requirement for the firms, enforced through the client governments, is that the contractors adhere to standards similar to those imposed on federal agencies in the United States by the National Environmental Policy Act. This means that the engineering and construction firms doing the actual work of building a hydroelectric system, reforesting millions of acres, or controlling insect pests are required to engage the services of environmental consultants before getting the funding to undertake the project. Then, throughout construction, they must follow the recommendations of these consultants. Environmental engineers often are the key members of these evaluation and approval teams.

What Do Environmental Engineers Earn?

Civil engineering—of which environmental engineering is a subspecialty—traditionally has been the lowest-paying of ten major branches of engineering. According to the College Placement Council, all engineering graduates with a bachelor's degree averaged about $34,000 a year in private industry in 1992, but C.E.s averaged only $29,376 to start, compared to over $40,000 for the highest-paid, the petroleum engineer.

A separate, 1993 salary survey of the engineering profession by the American Association of Engineering Societies reported a median starting salary for an environmental engineer of $33,350 for someone at the B.S. level. As customary, state government pays the least (among components surveyed), and large private companies pay the most, with salaries higher in metropolitan areas. Salaries for environmental engineers with ten years' experience generally are in the $45,000 to $55,000 range.

The AAEE's Buterbaugh writes: "The trend in salaries for both beginning and experienced environmental engineers indicates that demand is relatively weak. Staring salaries for those with a bachelor's degree, for example, averaged $25,000 five years ago, and $20,000 ten years ago. The salaries of environmental engineers today are the same as in 1970, before the 'environmental revolution,' in equivalent dollars adjusted for inflation."

What Is the Job Future for Environmental Engineers?

The Bureau of Labor Statistics expects the employment of civil engineers to increase about as fast as the average for all occupations through the year 2005, spurred by population growth and an expanding economy. More civil engineers will be needed to design and construct expanded or replacement transportation, water supply, and pollution control systems; contribute to large buildings or other structures; and repair or replace existing facilities. Most job openings, however, will result not from growth but from the need to replace an older generation of individuals leaving the work force for whatever reason.

The AAEE's Anderson sums up:

> Overall, long-term prospects for the environmental engineer are solid. There will always be work, despite the ups and downs of economic cycles. Like other types of engineering, environmental work offers modest monetary rewards. However, it can be an immensely satisfying and challenging career, which offers tremendous scope to make a difference. There are few things more fulfilling for someone with a technical bent than to combine a fascination with engineering with a professional commitment to

making the world a safer, healthier place. I believe environmental engineers have the best of all professional worlds.

ENVIRONMENTAL HEALTH SCIENCES

This is the science counterpart to environmental engineering. Its practitioner may have been educated in the life sciences, the physical sciences, or the social and behavioral sciences, and may have one or more degrees in ecology, biology, or chemistry. His or her graduate school education may be in agronomy, anatomy, animal science, bacteriology, biochemistry, botany, embryology, microbiology, pathology, pharmacology, physiology, zoology—the list could go on and on. The important distinction from the disciplinary scientist is that in the course of such a person's career, a new turn has been taken. It has been a turn toward the new environmentalism—*interdisciplinary/integrated/involved.*

What Do Environmental Health Scientists Do?

As an *environmental health scientist,* you would hold a key position on a team concerned with environmental protection or public health. You would be in command of the specialized knowledge, often scientific, connected with environmental problems. You would be aware of, and sensitive to, the concerns of others. You would assume command of the team when its assignments were largely scientific. Frequently you would: (a) search for, detect, and analyze environmental pollutants; (b) characterize and study the biological effects of pollutants (or of things such as noise which cause stress and are called "stressors"); (c) determine their epidemiological character (meaning their distribution geographically or through populations, as an epidemic disease might spread); and (d) devise criteria and standards for abating the pollutants or otherwise solving the environmental health problems.

Because this is such a broad category, and its practitioners frequently add management degrees to their scientific skills, you will find many environmental health scientists occupying middle-level or top administrative positions in which they run programs and supervise others.

What Education Do Environmental Health Scientists Need?

A bachelor's degree in biology, chemistry, or general science from literally any college or university would be an appropriate foundation for this field. If an institution offers an environmental science major, that too might be suitable if it heavily emphasizes science. Along the way, you should take as many management, statistics, and computer sciences courses as possible. You should aim, as soon as possible, for graduate school and get a master's degree. This might be a master's in public health administration, public administration, business administration, management, or one of the environmental sciences.

Where Do Environmental Health Scientists Work?

Local, state, and federal government agencies employ the majority of environmental health scientists, in departments such as public health, environmental protection, conservation, and fish and wildlife services.

The armed forces offer some of the more attractive career opportunities in environmental health. Colleges and universities employ environmental health scientists to teach and do research. Industry has a growing number of environmental management departments and research laboratories where such personnel are in demand—in part, to keep up with the flow of paper needed to comply with the tremendous number of government regulations regarding health and safety.

At the Centers for Disease Control and Prevention (CDC) in Atlanta, the Center for Environmental Health and Injury Control advertised for mathematicians/statisticians, pharmacologists, and environmental health scientists. For example, CDC sought members for its Emergency Response Coordination Group, which provides a 24-hour service for consultation when disaster strikes. On standby are physicians, toxicologists, environmental health scientists, chemists, health physicists, epidemiologists, and emergency response coordinators. A team can be assembled within 20 minutes and dispatched anywhere in the world, as one was when the earthquake struck northern California in 1989. In between such disasters, the CDC staff develops contingency plans and computerized systems and data bases that are needed in emergencies. This planning and program development takes place in conjunction with other agencies and the private sector.

The National Institute of Environmental Health Sciences—a component of the National Institutes of Health and headquartered at Research Triangle Park, North Carolina—recently issued an announcement that it was recruiting physician and nonphysician epidemiologists, health scientists, and microbiologists.

The National Institute for Occupational Safety and Health—a component of CDC located in Cincinnati, Ohio—frequently announces openings for environmental health scientists.

All the named government agencies, and others, have cooperative education, Presidential Management Intern, stay-in-school, and summer employment programs.

What Do Environmental Health Scientists Earn?

Current salary ranges for M.S.- and Ph.D.-holding candidates for such jobs—individuals with three to five years experience—at the GS-13/14 levels begin at approximately $50,000 per annum. (The figure includes "locality pay," a 5 percent or higher dividend in certain metropolitan areas with a high cost of living. There, federal workers are entitled to higher salary levels than civil servants employed elsewhere.)

Federal workers in the same localities with one to two years experience would qualify for GS-11/GS-12 jobs, with salaries beginning at approximately $30,000 per annum. Salaries in industry would be somewhat higher.

What Is the Job Future for Environmental Health Scientists?

Health science is a segment of the bioscience field, involving the application of biological science to human health needs. There is inadequate information, however, about the number of workers or future employment needs. An improving economy, reduced federal deficit, and increasing appropriations for health programs would strengthen the ability of governments, at all levels, to hire environmental health scientists.

As health science is a segment of the bioscience field, many of the statements made in Chapter 3 regarding the job future for biologists apply, as well, to environmental health scientists. This is particularly true with regard to the expected impact of biotechnology on job opportunities for biological scientists and environmental chemists. Similar assessments can be made, as well, for environmental health scientists. They are needed in the biotechnology industry—from the Department of Defense's biological warfare laboratories, to CDC's massive germ-containment laboratories, to industry's extensive clean-air compliance efforts.

According to *Healthy People 2000,* the Department of Health and Human Services forecasts that 121,000 additional environmental health specialists will be needed by the year 2000.

"Currently, there are only 1,500 environmental health graduates nationwide each year," according to the report. "As a result, government agencies are filling positions with inappropriately educated personnel .... Increasing the number of specialists in environmental health should be undertaken in conjunction with training in environmental health for physicians, generalists in medicine and engineering, and public school teachers."

The need is recognized; the barrier is in setting priorities and finding funding under tight budgetary constraints.

ENVIRONMENTAL MEDICINE

The category of *environmental medicine* consists of three different professionals whose functions are in some ways similar to those of environmental health scientists but differ in one respect: They are the only professionals who can treat patients. The degrees, with the abbreviations that distinguish them, are:

Doctor of Medicine (M.D.)
Doctor of Osteopathy (D.O.)
Doctor of Veterinary Medicine (D.V.M.)

Any one of the three professionals listed could be in charge of a particular environmental health team for disease control, laboratory research, or field studies; which one would be in charge would depend upon which individual was senior and had the expertise to deal with the problem confronted.

THE ENVIRONMENTAL PHYSICIAN

The National Association of Physicians for the Environment (NAPE) was established in 1992, suggested by an editorial simultaneously published in the *Journal of Environmental Health* and the *American Journal of Industrial Medicine.*

One medical organization, the American Academy of Otolaryngology—Head and Neck Surgery, Inc., was immediately so enthusiastic it asked each of its members to post a certificate pledging its member doctors and associated nurses and office staff "to help improve our environment by adhering to five principles." These range from leadership roles in the community and specific ways of making office practices ecologically sound to generally working "to protect biological diversity on earth from destruction by pollution and loss of habitat."

"Human health is inseparable from the health of the natural world," says NAPE executive vice president, John T. Grupenhoff, Ph.D. "In recent years, physicians increasingly have become interested [in] and concerned about our deteriorating environment, and the potential impacts on human health; physicians are the most widely distributed scientifically trained professionals in the United States."

Like some other professional organizations, NAPE is both a consortium of other groups and a society of individuals. All individuals, not only physicians, can join NAPE at a modest membership fee.

"Physicians are much more likely to involve themselves in environmental issues if [their] organizations are involved," explains Grupenhoff, if "they depend upon, and trust, these medical specialties for medical policy and development and continuing professional education and information. Therefore, from the beginning, NAPE also has been a consortium of medical groups."

Almost 100 medical organizations have become affiliated with the National Association of Physicians for the Environment. They include the American Medical Association, Aerospace Medical Association, Wilderness Medical Society, numerous specialty groups, and state and local medical societies.

With sponsorship of some of those societies and support from the National Institutes of Health—especially NIH's National Institute of Environmental Health Sciences—NAPE has held national conferences on air pollution's impacts on the body, biological diversity and human health, and other topics.

The American Medical Association (AMA) is concerned, on behalf of its members, all of whom are physicians, about a host of environmental health matters, and it is strongly supportive of NAPE.

Among the issues on which the AMA's House of Delegates has taken positions are agent orange/dioxin, AIDS/HIV virus, air pollution, alcohol and driving, asbestos, athletic training, automobile injuries, automobile safety, biological warfare programs, boxing, confidentiality of occupational medical records, drug use by athletes, drug testing of employees, emergency planning, employees' right to know (about workplace health hazards), energy health risks and nuclear energy, environmental stewardship, fire protection, flame-resistant apparel, fluoridation of public water supplies, formaldehyde, global climate change and the greenhouse effect, handguns, immunization, infant mortality, infectious medical waste, ionizing radiation, lead, ozone, pesticides, pollution control, radiation emergencies, radioactive wastes, radon, sanitary facilities for agricultural workers, smallpox vaccination, snuff and chewing tobacco, sports medicine, tobacco smoking, venereal (sexually-transmitted) disease, water pollution, and workers' reproduction rights.

What Do Environmental Physicians Do?

Many physicians emerge from the long course of education and training—eight to ten years or more beyond high school, the longest regimen of any profession—challenged by the opportunity not just to *treat* disease, but to *prevent* it. They may have been idealistic to begin with. Something then may have happened to push them in the direction of preventive medicine and public health. It could have been exposure to a charismatic professor, experience in working with the poor in a slum clinic, a stint in the uniformed services, an overseas internship, or an assignment in the Peace Corps or the Indian Health Service of the Public Health Service. Preventive medicine and public health (environmental health) is a recognized medical specialty.

What Education Do Environmental Physicians Need?

Basic education—four years of college and four years of medical school—is required of all physicians, regardless of later specialization. Specialty training may take an additional two to five years or more.

The high school student aiming for premed and med school takes about what any science major would take—math, science, English, speech, and social studies. In college, one continues with advanced courses in all those subjects, plus physics, basic calculus, biology, organic and inorganic chemistry, and one or two languages. The standard premed curriculum is preferred, with environmental science/environmental studies as part of available electives.

Medicine is *the* most difficult profession to enter. It enjoys the highest prestige and the highest level of remuneration of almost any calling, so it attracts the brightest students with outstanding secondary school records. Competition for limited slots in first-year classes is extremely keen. If you aim for a career in medicine, aim high and aim early. At about the junior year in college, you must take a medical college preadmissions test. (See your counselor about arrangements to take it, as well as about meeting the other qualifications for admission to medical school.)

The medical-school curriculum includes anatomy, pharmacology, biochemistry, physiology, microbiology, and pathology, plus clinical subjects such as pediatrics, radiology, obstetrics-gynecology, and internal medicine. Laboratory and clinical training are increasingly integrated throughout the four years.

All states require a license to practice medicine. Additional specialty training may be required. To conduct research or to administer programs, a license is not required, but many researchers and administrators with healing arts degrees choose to be licensed.

Environmental medicine is particularly rich in the resources available for lifelong learning. This specialist may choose, usually after a few years of practice, to attend a school of public health for the Master of Public Health (M.P.H.) or Doctor of Public Health (D.P.H.) degree. The purpose of this additional education might be primarily to acquire new scientific skills. It might be to become better qualified to teach public health methodology. Or it might be to learn to manage a complex program (something medical school does not teach). Of course it might be all three things together.

Where Do Environmental Physicians Work?

The environmental physician may be full-time, working on the staff of a corporation or public health agency, or part-time, maintaining a private practice in addition.

In a corporation, the function might be called occupational or industrial medicine. In a state or local health department, the physician might or might not see patients, administer shots, teach in a medical school, or lecture on health matters to school or community groups.

If part-time, the physician might spend a day or half-day a week operating a clinic, either general or specialized—for example, taking care of patients with acquired immunodeficiency syndrome, or AIDS—giving shots, testing children for high blood lead levels, or advising on the risks of waterborne disease or toxic agents. Part-time environmental physicians normally are internists or are in one of the internal medicine specialties—cardiology, infectious diseases, gastroenterology, and oncology, among others.

If full-time, the physician might practice clinical medicine to some extent, or might be completely occupied by managerial or administrative tasks. The post likely would be on the staff of a local or state health department; and in a large department its occupant would specialize in one program. This program might involve any of the diseases or specialties already mentioned, such as AIDS; women's, maternal, and child health; or infectious diseases.

At the federal level, this physician might be one of the renowned "disease detectives" of the Centers for Disease Control and Prevention in Atlanta. These investigators keep their bags packed, ready to fly on a moment's notice anywhere in the nation or the world to help quell mysterious disease outbreaks, which often are associated with the physical environment. Examples are legionnaire's disease and toxic shock syndrome. AIDS was first identified in 1981 by CDC, as was, later, the quickly deadly hantavirus, which emerged among members of the Navajo Nation in New Mexico in 1993. "Flesh-eating bacteria" startled everyone when it emerged around the world in 1994.

Some environmental physicians teach at schools of medicine or public health. Many retire to lucrative jobs in industrial medicine, pharmaceutical research or testing, clinical medicine, product development, toxicology, or other private-sector endeavors. Others expand, or take up, private practice, either in the places where they have worked or in Sun Belt retirement areas.

What Do Environmental Physicians Earn?

The *Occupational Outlook Handbook* states, "A physician's training is costly .... While education costs have increased, student financial assistance has not. Scholarships, while still available, have become harder to find. Loans are available, but subsidies to reduce interest rates are limited."

After the four years in medical school come three to six years of residency, for which allowances of $29,000 to $36,000 a year are common. Many hospitals also provide full or partial room and board, but the total is not munificent. Additional years are required for specialty training. Finally, there may be the high expense of buying or relocating a practice or opening an office. This financial investment will vary according to the equipment required by the specialty—relatively low for a psychiatrist, high for a radiologist.

All of this could add up to half a million dollars before the physician can even dream of a club membership or a Mercedes. During the first year or two of independent practice, many physicians barely pay expenses. As a rule, however, their earnings rise rapidly as their practice develops.

A new federal law permits the National Institutes of Health to pay a number of its top physician-scientists premium salaries of over $150,000 per annum. It remains to be seen how many key people this range of compensation will help retain when dozens have left in recent years citing the need for more money. Some are reputed to have gone to pharmaceutical and biotechnology companies for $500,000 or more, plus stock options and other benefits. Commonly, those who have departed for medical school deanships and endowed faculty posts are said to receive $300,000 to $400,000. It is difficult to generalize about salaries for physicians in public health, epidemiology, or environmental medicine, but fellowships at NIH and elsewhere are commonly advertised at $40,000 to $50,000.

What Is the Job Future for Environmental Physicians?

The Bureau of Labor Statistics expects overall employment of physicians—in private practice and other settings—to grow faster than the average for all occupations through the year 2005. However, reports the BLS: "Some health care analysts believe that there is, or that there soon could be, a general oversupply of physicians; others disagree."

Had national health care reform been enacted by Congress in 1994, when it considered but rejected several plans, all aspects of medical practice would have been affected.

Nevertheless, the BLS says:

Unlike their predecessors, newly trained physicians face radically different choices of where and how to practice. Many new physicians are likely to avoid solo practice and take salaried jobs in group medical practices, clinics, and health maintenance organizations (HMOs) in order to have regular work hours and the [enhanced] opportunity for peer consultation. Others will take salaried positions simply because they cannot afford the high costs of establishing a private practice while paying off student loans.

Health care reform advances, whether by deliberation or default. However, it is deliberate that the nation is devoting unprecedented attention to environmental and preventive medicine. In 1991 the National Academy of Sciences' Institute of Medicine (NAS-IOM) published a report indicating a need for between 3,100 and 5,500 additional physicians with special competence in occupational and environmental medicine.

In mid-1995 the same nongovernmental body, NAS-IOM, issued a comprehensive study on environmental medicine education. It proposed, for medical students, a new curriculum in environmental medicine, and incorporated more than 50 case histories of successful enhancement of environmental medicine studies.

PUBLIC HEALTH VETERINARY MEDICINE

Traditionally, the *public health veterinarian* has worked in a profession and in settings where it was necessary to be *interdisciplinary/integrated/involved* with society and the human environment. In taking the professional oath, says the American Veterinary Medical Association (AVMA), the veterinarian "solemnly swears to use her or his scientific knowledge and skills 'for the benefit of society, through the protection of animal health, the relief of animal suffering, the conservation of livestock resources, the promotion of public health, and the advancement of medical knowledge.'"

In a booklet entitled "Tracking Diseases from Nature to Man," published by the Centers for Disease Control and Prevention, we find:

People have only to conserve plant and animal life, and the resources of the earth, to share in nature's bounty and beauty. But they can also suffer because of this close relationship with nature: Many diseases pass from animals, insects, and soil to humans. The beautiful blue heron, like the golden-haired child with diptheria, can be the source of disease.

What Do Public Health Veterinarians Do?

Public health veterinarians are directly concerned with the promotion of public health and the advancement of knowledge, but that does not explain what they do. Simply put, what they do is intervene and interrupt the transmission of disease from animals to humans.

There are some 40 diseases to which humans are prone, in which animals play some role in transmission. Such a disease is called a *zoonosis,* defined by the American Public Health Association as "an infection or infectious disease transmissible under natural conditions from vertebrate animals to [humans]."

The most dreaded such disease is rabies because it is almost invariably fatal if not treated immediately. This acute viral infection most often is transmitted from rabid dogs, foxes, skunks, and raccoons, among other biting animals. Vampire and fruit-eating bats of Central and South America are infected with the rabies virus—but rarely are the source of human infections. When anyone is bitten by an animal suspected of being infected, the animal, if available, should be killed and examined for rabies. If rabies is determined—or if the animal cannot be caught for examination—the victim must be vaccinated as soon as possible after the bite, revaccinated four more times, then observed by a physician for at least 90 days.

Anthrax, to take another example, may be transmitted through the skin from contact with contaminated hair, wool, hides, and manufactured products containing those substances.

Certain types of food poisoning may be acquired by eating products containing frozen or dried eggs contaminated with salmonella. Q fever is an airborne infection sometimes suffered by veterinarians, dairy workers, and farmers. Psittacosis is called "parrot fever," and it strikes those who work around birds. Several forms of encephalitis can be transmitted by mosquitoes from horses to humans. Rocky Mountain spotted fever is transmitted by ticks. Trichinosis, whose infectious agent resides in swine, is transmitted through eating infected pork.

In the *Occupational Outlook Handbook* we read:

> Veterinarians help prevent the outbreak and spread of animal diseases, some of which—like rabies—can be transmitted to humans, and perform autopsies on diseased animals. Some specialize in epidemiology or animal pathology to control diseases transmitted through food animals and to deal with problems of

residues from herbicides, pesticides, and antibiotics in animals used for food.

In the newest edition of "Today's Veterinarian," a pamphlet issued by the American Veterinary Medical Association (AVMA), the range of public health veterinarians' duties is further extended:

> Veterinarians also work in the area of public health for city, county, state, and federal agencies. They help to prevent and control animal and human diseases and [to] promote good health. As epidemiologists, they investigate animal and human disease outbreaks such as food-borne illness, influenza, plague, rabies, AIDS, and encephalitis. They evaluate the safety of food-processing plants, restaurants, and water supplies. Veterinarians in environmental health programs study and evaluate the effects of various pesticides, industrial pollutants, and other contaminants on people as well as animals.

What Education Do Public Health Veterinarians Need?

There are 27 accredited veterinary colleges in the United States. Each prescribes its own preveterinary admission requirements. Typically, these include basic language arts, social sciences, humanities, mathematics, chemistry, and the biological and physical sciences. Three years of college is the minimum required for admission. The standard premed curriculum would be quite suitable; the environmental science/environmental studies curriculum would generally not be.

In most colleges of veterinary medicine, the professional program is divided into two phases. During the first phase, the student takes preclinical sciences including anatomy, physiology, pathology, pharmacology, and microbiology. Most of the time is spent in classroom and laboratory study.

The second phase of professional study is largely clinical. Students work with animals and deal with owners who use the school's clinical services. The clinical curriculum includes courses on infectious and noninfectious diseases, advanced pathology, applied anatomy, obstetrics, radiology, clinical medicine, and surgery. Applied studies include public health, preventive medicine, toxicology, nutrition, professional ethics, and business practices.

A veterinary degree program is rigorous, requiring about four thousand hours programmed in classroom, laboratory, and clinical study. Because of this heavy schedule, the student must devote many additional hours at night and on weekends and holidays to reading assignments, library research, and independent study.

The program leads, at the end of four years, to award of the Doctor of Veterinary Medicine (D.V.M.) degree.

Where Do Public Health Veterinarians Work?

The new veterinary graduate may qualify for a civilian job with the U.S. Department of Agriculture as a meat or poultry inspector, or for a commissioned corps appointment to the U.S. Public Health Service as an epidemiologist. Depending on the agency and the duties, a license from a state—any state—may be required.

The career public health veterinarian usually is employed by a unit of local, state, or federal government. As the basic veterinary medical education is rigorously scientific, many veterinarians become health administrators in programs remote from animal health. According to the Bureau of Labor Statistics, veterinarians held about 44,000 jobs in 1992—a rare declining profession. Most were in private practice. About 2,000 were employed by the federal government, chiefly in the Departments of Agriculture, Health and Human Services, and the Interior. An additional 600 served in the military.

Other important employers of veterinarians are colleges of veterinary medicine and medical schools, research laboratories, livestock farms, animal food companies, pharmaceutical firms, and biotechnology companies.

With the emergence in the late 1980s of intensified concern over the ethical and humane treatment of animals in biological research, the public health veterinarian has found a new role: caretaker of and advocate for animals who live in research laboratory animal colonies and are used for experiments.

When a zoonosis is even remotely possible, veterinary epidemiologists on the staff of, or in training at, the Centers for Disease Control and Prevention (CDC) are among the "disease detectives" who investigate the outbreak.

A veterinarian can meet military requirements or make an interesting career serving in the veterinary corps of the army or air force. One's functions in such service would usually relate to protection of the health of military personnel against the zoonoses described above. But duties also could be general sanitation, research, or a hundred other functions indistinguishable from the general-science duties which might be performed by a Ph.D. or M.D. A veterinarian also can meet military requirements by service in the commissioned corps of the Public Health Service, most commonly at the CDC, or at a state or local health department.

Veterinarians in the Science and Education Administration of the U.S. Department of Agriculture participate in a wide range of research activities: to improve livestock (and poultry) productivity through improved breeding, feeding, and management practices and to develop methods for controlling animal diseases, and parasites and insect pests which affect livestock. In addition to basic and applied animal research, these scientists have new priorities such as research to develop energy technologies to reduce animal agriculture's dependence on fossil fuels. Veterinarians, among others, are confronted by such questions as: Can manure from various animal operations produce enough methane gas to supply nearby buildings with heat and electricity?

Much research of that type is conducted in federal laboratories and in cooperative experiment stations of the state colleges and universities. It also is conducted by other university staffs under federal contracts and grants. Hundreds of veterinarians, federal and otherwise, participate in these studies.

What Do Public Health Veterinarians Earn?

In 1994, according to data published by the AVMA, the average starting salary of veterinary medical college graduates was about $29,000 per annum, but sometimes reached as high as $33,000. Those in private practice, especially those who own their own veterinary hospitals, earn considerably more than the estimated $60,000 to $70,000 maximum for most public health veterinarians.

What Is the Job Future for Public Health Veterinarians?

Employment of veterinarians is expected to grow faster than the average for all occupations through the year 2005, according to the Bureau of Labor Statistics. The prediction is hard to rationalize when, as noted, the *Occupational Outlook Handbook,* 1994–1995 edition, shows veterinarians held about 44,000 jobs in 1992, or *2,000 fewer than in 1990!* More likely correct is the American Veterinary Medical Association's 1994 estimate for the total number in practice: "Today over 55,000 veterinarians are professionally active in the United States."

BLS presents its explanation of employment growth as follows:

> The number of pets is expected to show a steady increase because of [owners'] rising incomes and the movement of baby boomers into the 34–69-year age group, for which pet ownership is highest. Pet owners may also be more willing to pay for more intensive [pet] care than in the past. In addition, emphasis on scientific methods of breeding and raising livestock and poultry, and continued support for public health and disease control programs will contribute to the demand for veterinarians.

ENVIRONMENTAL SANITATION

If you are more interested in practical problems than in theory, better satisfied by helping people avoid trouble than by working at a laboratory bench, then consider becoming a *sanitarian.* (The sanitarian is not to be confused with the sanitary engineer, which is what environmental engineers traditionally were called, and whose duties have been described in earlier pages.)

What Do Sanitarians Do?

The sanitarian traditionally has been responsible for interpreting and enforcing local, state, and federal laws, regulations, and standards—mostly local—respecting the sanitation of food, milk, water supply, garbage disposal, sewage disposal, and housing. Today, to those responsibilities has been added the stimulation of community action for better health through better environmental sanitation of all types. This means that the sanitarian may promote and help secure such improvements as water supply extension, improved sewage disposal facilities, safer recreational areas, and more hygienic conditions in nursing and convalescent homes.

One level of duty is that of *inspector,* checking on the cleanliness of dairies, food-processing plants, restaurants, and plant and hospital food services. As a sanitarian you would also visit schools, hotels and motels, swimming pools, parks, and certain types of housing to observe, make measurements, collect samples and specimens for laboratory analysis, and make recommendations regarding the facilities' compliance with public health and environmental protection and safety laws and regulations. Another level of duty is that of *regulator,* citing violators of codes or even closing them down.

A *supervising sanitarian* plans and directs environmental health programs and may supervise a large staff. Sanitarians at all levels serve the public as educators and interpreters. They provide consultation to owners, operators, and employees of businesses; to contractors; to school, government, and elected officials; and to any citizens who wish to pose questions or make requests. They may draft proposed laws and regulations. They may give expert testimony in court cases involving alleged violations of such laws and regulations. They may serve as arbitrators in quasi-judicial proceedings. With the recent emergence of new and more complex pollution problems, especially those involving toxic wastes and hazardous substances, the sanitarian can expect to be called on far more frequently, by all the parties named, for expert advice and counsel.

What Education Do Sanitarians Need?

A bachelor's degree in biology, chemistry, general engineering, or environmental sciences generally is acceptable for an entry-level job as sanitarian. This could be an environmental science/environmental studies degree providing it is strong in science electives. In most states one cannot practice as a professional sanitarian without a bachelor's degree. More than 60 colleges and universities offer a bachelor's degree in environmental health (or sanitation), which requires a minimum of 30 semester hours in the physical or biological sciences. The communications and liberal arts courses mentioned so frequently in connection with other professions are stressed. A typical curriculum's core courses would include microbiology (bacteriology), biostatistics, epidemiology, environmental sciences, administration, and field work.

A graduate degree is not offered in sanitation as such, but sanitarians frequently take graduate work, through the master's level and sometimes the doctorate, in science, public health, management, or some other field.

The National Environmental Health Association (NEHA) awards an option of three credentials to members who meet educational and experiential requirements and may need to complete examinations. They are (a) Certified Hazardous Waste Specialist (CHWS), developed under a grant from the Environmental Protection Agency, which requires a bachelor's degree with major course work in the sciences or in environmental health; (b) Registered Environmental Health Specialist (REHS), required by state registration boards, which has as prerequisites a bachelor's degree and a minimum of two years' environmental health experience; and (c) Environmental Health Technician (EHT), which is in increasing demand for positions in the military and which requires a high school diploma and a minimum of two years' environmental health experience or an associate's degree in a related field.

NEHA, with support from the Environmental Protection Agency and the Public Health Service, has developed 17 self-paced learning modules for the individual's continuing education. These modules cover such topics as water quality, wastewater treatment, air pollution, injury and disease prevention, noise control, data management, and communications.

Where Do Sanitarians Work?

Three out of four sanitarians work for the federal, state, or local governments. Another group works for producers and processors of food and dairy products. Others teach, consult, or work for hospitals, trade associations, or such firms as insurance companies. Because of their broad background in environmental sciences, such people readily move back and forth between the private and public sectors. With added educational and experiential qualifications, they may seek steady advancement and new challenges.

What Do Sanitarians Earn?

According to the Bureau of Labor Statistics, health and regulatory inspectors, including sanitarians, start at around $18,000. That would be equivalent to a GS-5 grade, where step 1 carries a beginning salary of $18,500.

The chances for advancement are good, and by midcareer one could easily be making $50,000 to $60,000 per year or even more.

Government sanitarians come into contact with owners of businesses and executives of such companies as food processors. They may be offered jobs by these contacts, and in changing jobs improve their salaries by 50 percent or more.

What Is the Job Future for Sanitarians?

At the same time that the duties of the sanitarian have broadened, the number of businesses which must be inspected has expanded. Think of the rapid growth that you have observed of restaurants, shopping malls, and fast-food places.

The emerging genetic engineering and biotechnology industries have been mentioned previously, in Chapters 3 and 4. Some research in these areas requires operations to be conducted in a germ-free environment. For example, the Centers for Disease Control and Prevention, at its headquarters in Atlanta, has a large germ-free containment building for safely storing and studying the agents of anthrax, rabies, smallpox, and other deadly infectious agents. Sanitarians are among the specialists who helped design and who operate the facility.

As the BLS does not include sanitarians as a category in the *Occupational Outlook Handbook,* there are no "official" projections as to the job future for sanitarians. It can be assumed that employment of sanitarians will grow about as fast as the average for all occupations through the year 2005. Growth should be fueled by the expected surge in biotechnology in agriculture and the food industries, continuing growth of the population and the economy, and the expected expansion of radiation sterilization of foods.

INDUSTRIAL HYGIENE

What Do Industrial Hygienists Do?

Many specialists are trying to anticipate the potential cancer, birth defects, mental retardation, emphysema, black lung disease, and mental illnesses which might result in the years ahead from stresses and impacts in the workplace today. The *industrial hygienist* is the key professional who can take immediate, practical steps to identify problems, tighten standards, remove hazards, and ban toxic substances whenever they are identified. These steps alleviate short-term illness and contribute to long-term worker health.

Become an *industrial hygienist* and, as the title indicates, you would work in or with industry. You would conduct activities to eliminate if possible, or at least to control, occupational health hazards and diseases. Concerns include: (a) chemical stresses such as dust or gas, (b) physical stresses such as radiation or noise, (c) biological factors including insects and fungi, and (d) ergonomic (work-related) stresses such as monotony and work pressure.

In its educational booklet "Balancing Work, Health, Technology & Environment," the American Industrial Hygiene Association (AIHA) defines the field:

> Industrial hygiene is the anticipation, recognition, evaluation, and control of workplace environmental factors that may affect

the health, comfort, or productivity of the worker. Industrial hygiene is considered a "science"; however, it is also an art that involves judgment, creativity, and human interaction.

What Education Do Industrial Hygienists Need?

As explained in the 1994 edition of AIHA's educational booklet, few colleges or universities offer an industrial hygiene major leading to the baccalaureate degree. Industrial hygienists generally pursue an undergraduate degree in one of the sciences—such as engineering, chemistry, biology—or in industrial management. Just over 100 U.S. and Canadian universities offer B.S. or postgraduate degrees relating to industrial hygiene.

Some institutions offer a one- to three-year associate degree and a certificate program that qualifies students as industrial hygiene technicians. An environmental science/environmental studies bachelor's might qualify someone who was willing to start at the technician level.

After you have worked in industrial hygiene for five years, you are eligible for certification. To become certified, you must take a comprehensive two-day examination and meet other requirements. The certified industrial hygienist, or "CIH," confers professional benefits; for example, in EPA environmental regulations, the Environmental Protection Agency recognizes the special competency of CIHs. Often the industrial hygienist also gets a master of public health degree, or a law degree, either of which could be professionally beneficial.

Where Do Industrial Hygienists Work?

In 1994 AIHA surveyed 2,500 members in a representative sample and received a 63 percent response—1,575 of its then-11,000 members. As AIHA estimates there are 20,000 to 25,000 total in the profession, it feels it can "relate these figures comfortably to the entire profession."

It found 42 percent of industrial hygienists work for private industry, 19.6 percent for consulting firms, 13.6 percent for government, about 5 percent in insurance, and the rest who could be identified teach or are self-employed.

Since 1970, when Congress passed the Occupational Safety and Health Act, two government agencies have played clearly delineated roles. The Occupation Safety and health Administration (OSHA), which is part of the U.S. Department of Labor, is responsible for setting and enforcing health and safety standards across the nation. OSHA is headquartered in Washington, D.C. In addition, some special OSHA-approved offices that are administered by individual states also conduct health and safety inspections.

The other agency is the National Institute for Occupational Safety and Health (NIOSH), based in Cincinnati, Ohio, and a component of the Centers for Disease Control and Prevention (CDC). NIOSH was created as a research agency, to learn more about occupational illnesses and how to prevent them. "Some of the most exciting research in the field of industrial hygiene is performed at NIOSH," says AIHA. "NIOSH also aids in the training of industrial hygienists. A large number of industrial hygienists who received master's degrees since 1970 earned [them through] graduate scholarships paid for by NIOSH."

In 1991 the Department of Energy designated industrial hygiene as a "shortage occupation," which means that it is placed on a list of jobs whose practitioners qualify for higher-than-average civil service pay rates. The Energy Department also reported a 200 percent increase in its own employment of industrial hygienists. And the same year, the Energy Department also surveyed other agencies and reported that the number of industrial hygienists employed full-time in federal civil service in all government agencies grew 44 percent between 1985 and 1991. The fastest growth rate, surprisingly, occurred in the Department of Veterans Affairs (VA), which reported a tremendous 765 percent increase in industrial hygiene employment in those six years!

What Do Industrial Hygienists Earn?

According to AIHA literature, entry-level industrial hygienists in 1994 were generally paid between $28,000 and $35,000.

In its member survey, AIHA found only 10.5 percent of its members in that range, however; which was not surprising, as only 8.3 percent of the respondents reported fewer than five years experience. A whopping

15.4 percent—primarily in consulting for executive positions—earned more than $80,000 in 1993; another 41.8 percent were paid between $40,000 and $60,000.

What Is the Job Future for Industrial Hygienists?

"Industrial hygienists say the need for people in their specialty is so great that one does not need to have a Ph.D. to be a high earner," wrote Edward R. Silverman in reporting on the AIHA survey in *The Scientist* early in 1995.

AIHA puts it this way: "The need for competent industrial hygiene professionals is growing rapidly, and the demand is sure to increase even more in the coming years. In fact, 1992 and 1993 *U.S. News & World Report* surveys ranked industrial hygiene as one of the top professions of the future."

AIHA is believed to be the only professional organization in the environmental field that can claim a virtual *doubling* of its membership in just ten years! In 1983, AIHA had 5,954 members on its rolls; by 1993, membership had increased to 11,327.

WHERE SHOULD YOU WRITE FOR MORE INFORMATION?

Accreditation Board for Engineering and Technology
Student Services Department
345 East 47th Street
New York, N.Y. 10017

American Academy of Environmental Engineers
130 Holiday Court, Suite 100
Annapolis, Md. 21401

American Industrial Hygiene Association
2700 Prosperity Avenue, Suite 250
Fairfax, Va. 22031

American Public Health Association
1015 15th Street, N.W.
Washington, D.C. 20005

American Society of Civil Engineers
  Student Services Department
  345 47th Street
  New York, N.Y. 10017

American Veterinary Medical Association
  1931 North Meacham Road, Suite 100
  Schaumburg, Ill. 60196

Centers for Disease Control and Prevention
  National Center for Environmental Health
  (or National Institute of Occupational Health and Safety)
  1600 Clifton Road, N.E.
  Atlanta, Ga. 30333

National Association of Environmental Professionals
  5165 MacArthur Boulevard, N.W.
  Washington, D.C. 20016

National Association of Physicians for the Environment
  6410 Rockledge Drive, Suite 412
  Bethesda, Md. 20817

National Environmental Health Association
  720 South Colorado Boulevard
  Denver, Colo. 80222

National Institute of Environmental Health Sciences
  Office of Communications
  P.O. Box 12233
  Research Triangle Park, N.C. 27709

National Safety Council
  Environmental Health Center
  1050 17th Street, N.W., Suite 770
  Washington, D.C. 20036

Occupational Safety and Health Administration
  U.S. Department of Labor
  200 Constitution Avenue, N.W.
  Washington, D.C. 20210

Office of Disease Prevention and Health Promotion
  Public Health Service
  330 C Street, S.W., Room 2132
  Washington, D.C. 20201

CHAPTER 5

AGRICULTURE AND NATURAL RESOURCES

The forest resources of our country are already seriously depleted. They can be renewed and maintained only by the co-operation of the forester and the practical man of business.—President Theodore Roosevelt in 1903.

Roosevelt sponsored conservation not so much to preserve a domain for agriculture as to preserve and enhance the strength of the whole nation.—*The Republican Roosevelt,* John Morton Blum, 1954.

PROFESSIONS COVERED

Agricultural Sciences, Fisheries Conservation, Forestry, Range Management, Soil Conservation, Wildlife Management

President Theodore Roosevelt chose as the first chief forester of the nation Gifford Pinchot (1865–1946). In so doing, the president acknowledged the need for professionals in high positions to make conservation policy and to administer conservation programs. Pinchot, you see, was the first American to hold a degree in forestry. His philosophy was "Conservation means the wise *use* of the earth and its resources." (Italics added.) Poet John Muir (1838–1914), on the other hand, believed with Henry David Thoreau (1817–1860) that, "In Wildness is the *preservation* of the World." Thus, Pinchot and Muir became the symbolic leaders of opposing camps—users vs. preservers—which have been engaged in frequent conflict since the twentieth century began.

A notable early environmental controversy arose in 1901. It concerned development of a water supply for the city of San Francisco, which proposed to create a reservoir in the spectacularly scenic Hetch-Hetchy Valley in Yosemite National Park. Other sites were available, at higher cost. The question was whether an artificial impoundment should be allowed within a national park simply because it was more economical to build on that site than on others available. (The environmental impact statement process was designed, 70 years later, to weigh alternatives like this.)

Pitted on opposite sides of the controversy were Muir, the founder of the Sierra Club, who fervently believed in wilderness inviolate, and Pinchot, the first professional forester in America, who favored the Hetch-Hetchy Reservoir as a sensible use of resources. The economic argument carried the day, and Congress in 1913 approved the reservoir plan.

THE PROGRESSIVE MOVEMENT TO THE NEW DEAL

The Progressive Movement led by President "Teddy" Roosevelt during his two terms, 1901–1909, introduced the first official concern for conservation into public policy making. Roosevelt was both prodded and publicized by the reform journalists, called "muckrakers," and was guided and assisted in conservation matters by Pinchot, his chief forester. Together, they started two dozen irrigation or reclamation projects and added more than 125 million acres to the national forest system. President Roosevelt also sponsored a White House Conference on Conservation in 1909 to which he sent invitations, personally delivered by Pinchot, to the Prime Minister of Canada and the President of Mexico.

Bird-watchers, fishers, hikers, hunters, naturalists, and others who loved the outdoors—joined by scientists—developed the concept of wildlife sanctuaries. They convinced legislators to institute hunting and fishing licenses to help pay for wildlife protection. And they worked zealously to protect local forests, marshes, and mountains from destruction. Except for their little-known efforts, the Grand Canyon, Yellowstone's geysers, the California giant sequoias, Florida's Everglades, Georgia's Golden Isles, and Massachusetts's Cape Cod, among hundreds of other natural and historic treasures, could have been destroyed by now.

Another President Roosevelt—Franklin Delano Roosevelt, despite the serious distraction of World War II—made conservation of the land a keystone of his four terms. Between 1933 and 1942, more than two million young men, most of them previously unemployed, worked in the Civilian Conservation Corps (CCC) on a variety of outdoor restoration projects. They planted windbreaks on the Great Plains, started tree farms in the southeast, and hewed firebreaks across the Green Mountains of Vermont and the Sierra Nevadas of California.

BEHIND THE MODERN ENVIRONMENTAL MOVEMENT

After World War II, Americans were more conscious of the beauty and value of the land than ever before. In the 1940s and 1950s, several popular books brought environmental issues to widespread attention. Occasionally citizens mobilized to save and protect an environmental treasure. In the early 1950s, the Bureau of Reclamation decided that a dam should be built on the Green River at Echo Park, part of the Dinosaur National Monument in Colorado and Utah—Hetch-Hetchy all over again. This time the results were different. Wilderness preservationists, defensive about the sanctity of the national park and monument system, conducted a vigorous and successful campaign which brought the development to a halt in 1956.

During the 1950s, public demand for outdoor recreation was accelerating, and conflicts over finite recreation resources increasing. In 1958 Congress responded by establishing a commission to plan to meet recreation needs in America over the next 40 years. President John F. Kennedy received its report in 1962, and his successor, President Lyndon B. Johnson, implemented many of its recommendations. Prodded by his wife, Lady Bird, Johnson viewed the environment as embracing beauty as well as utility. In 1965 he sponsored a White House Conference on the order of Teddy Roosevelt's conference of 1909. An unprecedented wave of citizen action followed in the late sixties.

Rachel Carson's *Silent Spring* (1962) was mentioned in Chapter 1. An arm of the White House established an expert committee on pesticides to scientifically study the issues Carson had raised, and its report largely vindicated her by recommending that many of her suggestions be made national policy.

On March 18, 1967, a tanker ran aground on Seven Stone Reef off England's southern coast. The resulting oil spill inflicted great damage on marine life and on the beaches of England and France.

That accident in Europe caused the U.S. government to make contingency plans to cope with the spillage of oil or hazardous substances off its shores. Those plans proved invaluable when, on January 28, 1969, an oil well in the Santa Barbara, California, harbor blew, discharging enormous quantities of oil into the Pacific Ocean. Within three days oil covered nearly 200 square miles of ocean and began washing up on beaches for hundreds of miles. Without all types of communications media, especially color television, the impact on public opinion in this country might have been negligible. Millions of readers and viewers, however, were emotionally moved by pictures of thousands of seabirds struggling to survive their coatings of oil and dispersing chemicals and were struck by the valiant efforts of volunteers trying to clean and save them. (Almost 20 years later, in 1988, an even greater spill followed the grounding, in Alaskan waters, of the *Exxon Valdez.* There followed the most extensive environmental cleanup ever attempted.)

It is ironic that oil spills still regularly occur, particularly because Santa Barbara was the symbolic trigger for the first Earth Day, which launched the modern environmental movement that now has reached worldwide proportions.

In this chapter we are concerned with natural resources, their protection, enjoyment, and diverse uses, from food and fiber to forests and fishes.

AGRICULTURAL SCIENCES

American agriculture is the most efficient in the world. Per capita, it is the most productive segment of our economy. Credit for American agriculture's accomplishments goes largely to the land-grant colleges and universities with their network of cooperative state research and extension services. (There is no denying the contributions of technology and economies of scale, as small farms have been consolidated into more efficient large farming operations. In this respect, agriculture is similar to the industrial and marketing sectors, where expansion, mergers and absentee ownership characterize the business world.)

American agriculture is changing in response to changing social needs and demands. Its overdependence on chemicals is being questioned by environmentalists, and many farmers are moving into integrated pest control, using insecticides as little as possible. Early in the 1980s, organic farming was being promoted, for the first time, by the U.S. Department of Agriculture, and thousands of farmers were adopting its principles. Farmers were widely adopting new techniques of energy conservation in agriculture, at the same time making a growing contribution to energy supply through increased production of grain for alcohol that can be combined with gasoline to produce "gasohol."

The Bureau of Labor Statistics estimated that *agricultural scientists* held about 29,000 jobs in 1992, not counting those who taught. As combination teaching-research jobs in the land-grant colleges and universities and associated agricultural experiment stations form a very large component of the agricultural science community, omitting them from the census is a serious omission. Until it is corrected, we can only guess that there must be at least 10,000 teacher-researchers on those campuses in America.

The U.S. Department of Agriculture's Science and Education Administration operates eight major research centers and more than 150 other research facilities, located in virtually all the states. In addition, it sponsors research in hundreds of colleges, especially in the 74 land-grand agricultural colleges.

Federally sponsored research and development embrace conservation and management of such natural resources as soil, water, and forests; animal and crop production-protection; crop utilization and postharvest technology; human nutrition and family-resource management; and domestic and export marketing.

What Do Agricultural Scientists Do?

Become an agricultural scientist and you might study the effects of changing climate and weather on agricultural production; work to eliminate the sources of pollution that result from agricultural practices; explore uses of fuel made from plant products; search for ways to reduce the use of fossil fuels in producing crops and livestock; test the possibilities of aquaculture (farming systems for plants and animals that live in water); find safe, feasible ways to use organic wastes; explore biological (nonchemical) pest control methods; promote urban and agricultural integrated pest management practices; develop methods for assuring the safety and quality of food and food products; conduct basic and applied research in the animal and plant sciences; do controlled experiments in human nutrition and diet; develop community resources in rural areas; conduct home audits for energy conservation; or develop systems to assist the public in coping with natural disaster emergencies.

The agricultural sciences, like the biological sciences, form a broad and complex field. It is divided into many categories, the major ones being outlined but not detailed here. The *agricultural chemist* develops chemical compounds to control insects, weeds, fungi, and rodents. The *agricultural economist* deals with the production, pricing, and marketing of farm products. And the *agricultural engineer* designs agricultural machinery and develops methods to improve the production, processing, and distribution of farm products.

The *agronomist* experiments with field crops, and a subspecialist, the *horticulturist,* does the same with orchard and garden plants. The *animal physiologist* is concerned with livestock and the various parts of the animals—a subspecialist to the *animal scientist,* who is concerned with the breeding and management of farm animals. The *veterinarian* (already discussed in Chapter 4) is responsible for the health of farm animals. And the *entomologist* concentrates on the control of insects that may injure plants and animals.

Those are only a few of the specialists who might work on problems and issues of agriculture.

What Education Do Agricultural Scientists Need?

Many agricultural science professions can be entered with only the bachelor's degree, sometimes with a major in environmental science/

environmental studies, especially one from a college of agriculture and natural resources.

One of the fastest-growing areas of higher education—as a direct consequence of the environmental movement—is colleges of agriculture and natural resources. During the 1980s, when college enrollments generally were dropping and many schools began recruiting students, the nation's agricultural colleges had all the students they could accommodate.

A bachelor's degree in agricultural science is sufficient for some jobs in applied research or assisting in basic research, according to the Bureau of Labor Statistics. Degrees in sciences such as biology, chemistry, or physics—or in related engineering specialties—also may qualify some persons for some technical jobs. Very often, a bachelor's in environmental science/environmental studies would be suitable for an entry-level job.

Every state has at least one land-grant college or university which offers agricultural science degrees. The BLS says:

> A typical undergraduate agriculture science curriculum includes communications, economics, business, and physical and life science courses, in addition to a wide variety of technical agricultural science courses. For prospective animal scientists, these technical courses might include animal breeding, reproductive physiology, nutrition, and meats and muscle biology; students preparing as food scientists take courses such as food chemistry, food analysis, food microbiology, and food-processing operations; and those preparing as crop or soil scientists take courses in plant pathology, soil chemistry, entomology, plant physiology, and biochemistry, among others.

Where Do Agricultural Scientists Work?

The agricultural scientist might find employment literally anywhere, from Park Avenue in New York to a village in the Philippines. Major employers include the U.S. Department of Agriculture, state departments of agriculture and natural resources, colleges and universities and their state experiment stations, agribusiness, international agencies, engineering and consulting firms operating around the world, United Nations agencies, and foreign governments with their equivalents of any of the above.

The economics of some areas of technology make them feasible in developing countries but not in the developed nations. The desalinization of saline water, for example, is not yet economical in the United States. But in Saudi Arabia, which has far more oil than water, an extraordinarily expensive desalting program now meets most of the country's domestic water needs. Giant desalinization plants on the Red Sea process 500 billion gallons of water a day, which is then piped to major cities and agricultural areas hundreds of miles away. Plans are being made to serve every city in Saudi Arabia, and experiments are under way with solar processing systems.

What Do Agricultural Scientists Earn?

According to the College Placement Council, beginning salary offers for agricultural scientists with the bachelor's degree averaged $20,000 (for animal scientists) to $22,000 per year (for plant scientists) in 1992. This range was, discouragingly, no improvement over reports from four years earlier. Nor is there evidence that today the agricultural colleges are being deluged with students, as had been the case four or five years before. The supply of both students and jobs appears to have abated.

As most beginners in agricultural sciences and natural resources professions start out working for government, the civil service pay scale sets the standard. In 1994, the GS-5 salary scale began at $19,116 per annum, the GS-7 at $23,678, the levels at which most B.S. graduates begin. With the master's degree and/or several years' experience, one might be graded at GS-9, $28,964; GS-11, $35,045; or even higher.

Average federal salaries for experienced employees in these areas in 1993 were somewhat more encouraging, again led by the animal science specialists at $55,631, about $10,000 per annum more than the plant scientist.

As many of the federal jobs are at agricultural experiment stations affiliated with state land-grant colleges and universities, the opportunity usually is afforded to take graduate work, sometimes at reduced tuition.

What Is the Job Future for Agricultural Scientists?

Employment of agricultural scientists is expected to grow about as fast as the average for all occupations through the year 2005. The Bureau of Labor Statistics is not as optimistic as it was in the late 1980s, when it reported "the nation's agricultural colleges had all the students they could accommodate," and employment of agricultural scientists was "expected to grow faster than average for all occupations through the 1990s."

In the more recent *Occupational Outlook Handbook,* we find:

> Although enrollments in agricultural science programs have begun to increase again after declining for several years during the 1980s, opportunities should still be available in most major subfields of agricultural science. Animal and plant scientists with a background in molecular biology, microbiology, genetics, or biotechnology; soil scientists with an interest in the environment; and food technologists may find the best [employment] opportunities [among agricultural science graduates].

Bachelor's degree holders may find employment with the Soil Conservation Service, U.S. Department of Agriculture, or as salespersons or technicians with businesses that deal with ranchers and farmers—such as feed, fertilizer, seed, and farm equipment manufacturers. Other possibilities are working as cooperative extension service agents, agricultural products inspectors, and purchasing or sales agents for agricultural commodities or farm suppliers.

Future farm bills are expected to contain support for agricultural research in *sustainable agriculture.* This is a new term for a concept warmly embraced by environmentalists because it advocates all possible reduction of the use of chemical pesticides and herbicides, fertilizers, and toxic natural materials commonly used in agricultural production.

Another desirable objective of sustainable agriculture is the use of low-input farm management to enhance agricultural productivity, profitability, and competitiveness. There are ways to make progress involving new combinations of old conservation techniques focused on soil, water, energy, natural resources, and fish and wildlife habitat.

Organic farming and gardening are low-tech and environmentally friendly. "Organic" may have less potential than "sustainable," but they have much in common. Sustainable agriculture has the potential for large-scale applications with resulting economies of scale, which should make American agriculture ever more productive and competitive on the world market.

A countertrend, from the environmental viewpoint, may be biotechnology's incursions into agriculture. After ten years of effort, Monsanto persuaded the Food and Drug Administration to approve the marketing of the beef hormone known as BST, the first genetically engineered product to be used to increase food production. Opponents of all bioengineered products attribute bovine deaths and health problems in dairy herds to the new drug.

Agricultural scientists from many disciplines are working to alleviate potential health and environmental hazards, if any.

FISHERIES CONSERVATION

Millions of people who live along the shores of the earth's oceans, seas, bays, lakes, rivers, and streams depend on fish for food, perhaps as their only source of life-supporting protein. In the United States alone, in 1988, commercial fishers harvested five million metric tons of finfish and shellfish products worth $4 billion. Aquaculture (fish farming) of fishes and invertebrates in ponds and other closed systems has grown rapidly in the last two decades. Aquacultural production in North America is roughly 360,000 metric tons—mostly catfish, crawfish, trout, and salmon—valued at $600 million.

Aquaculture is a fascinating career, according to the American Fisheries Society (AFS). So is the management of recreational fisheries. Next to swimming, sport fishing is the most popular outdoor activity in North America. In 1985 U.S. anglers numbered 59 million, or nearly 27 percent of the population. Anglers fished a total of 988 million days and spent over $28 billion on licenses, tackle, food, lodging, boats, motors, transportation, and fuel.

Yet, adds the AFS:

Our fisheries resources need help. The demands and stresses that have been placed on many fisheries continue to threaten their productivity. Dredging, dam building, and shoreline erosion physically alter aquatic habitats and can kill fish and invertebrates or interfere with reproduction. Withdrawal of water from lakes and streams for domestic, industrial, and agricultural purposes also reduces available habitat. Release of pollutants into the water threatens survival of all aquatic organisms. Overfishing also threatens many fisheries, and competition for various species of fish by sport, commercial, and subsistence fishermen often leads to conflicts and complicates management.

Many of the same conflicts and complications pertain in Canada. So do many of the same environmental resource opportunities: "If you are planning to work near water, better check the fisheries act first—and avoid problems later," says an advisory on Canada's fish habitat law. Canada's Atlantic groundfish management plan is being revised to support the conservation of certain species in various fisheries. The federal government rejects British Columbia's claim that the national government is liable for loss of fisheries because an aluminum project harmed fish habitats. A new seal management plan is being considered. A 40-page guide to federal aquaculture programs has been issued. And a new brochure on careers in Canada's Ministry of Fisheries and Oceans—"A Sea of Opportunities"—has been published.

Canada and the United States share the Great Lakes region, and the region's problems, too. Since 1989, when the two countries launched the Great Lakes Action Plan, automakers on both sides of the border became committed to reducing and/or eliminating discharges of 65 pollutants; the walleye population in Nipigon Bay has been restored to 25 percent of its historic level, and a full recovery is expected; the bald eagle has returned to Lake Erie's shores; and new, low-cost sewage treatment techniques are being implemented that will save millions of dollars in treatment plant upgrades.

The American Fisheries Society has established a certification program as a means of setting guidelines for professional recognition.

Specified are 30 semester hours of biological sciences, including 4 aquatics courses with at least 2 in fisheries; 15 semester hours in physical sciences (chemistry and physics); and 6 each in mathematics/statistics and communications.

What Do Fisheries Conservationists Do?

The professional directly concerned with sports and commercial fisheries is the *fisheries conservationist* or *wildlife biologist* specializing in this area. This specialist studies the life history, habits, classifications, and economic relations of aquatic organisms. The science is an applied field, characterized by practical applications of biological sciences, especially managing fish hatcheries, conducting information and education programs for those in sports and commercial fishing, inspecting and grading fishery products for human and animal consumption, and measuring and promoting the market for fresh fish and processed fisheries products.

What Education Do Fisheries Conservationists Need?

As in biology and ecology, fisheries conservation occupations require a good high school education, including physics, chemistry, biology, English, communications, mathematics, and a foreign language. Some community colleges offer the associate's degree in fisheries. You can get good preparation at almost any four-year college or university by taking a degree in biology or zoology. Study, for starters, the principles of comparative anatomy, microbiology, genetics, chemistry, mathematics, computer programming, and statistics. Continue to hone your skills in communications, English, and the humanities.

If you prefer a specialized curriculum, seek out one of the 167 campuses representing about 150 colleges or universities offering specialized programs in fisheries science or conservation—including 15 in Canada and one in Mexico. Many institutions offer an M.S. as well as a B.S. program. If you plan to teach and do research, a Ph.D. is mandatory.

Where Do Fisheries Conservationists Work?

A majority of fisheries conservationists work for the federal, state, provincial, and territorial government agencies of the United States and Canada. Seven agencies of the U.S. Department of the Interior, notably the Fish and Wildlife Service, constitute the largest employer of fisheries conservationists. The National Oceanic and Atmospheric Administration of the Department of Commerce, the Environmental Protection Agency, and components of the Department of Agriculture also employ considerable numbers. Among other significant employers are about two dozen cooperative fishery units such as hatcheries attached to agricultural experiment stations. A few international agencies, such as the Food and Agriculture Organization (FAO) of the United Nations, hire fisheries scientists. Others are employed by private industries to develop food products, compile data, prepare environmental impact statements, or manage aquatic properties. A few are hired by environmental consulting firms, forest product companies, private clubs, and organizations to do a variety of work from stream management to public relations.

What Do Fisheries Conservationists Earn?

As civil service salaries for federal employees are standardized, fisheries conservationists in the Department of Agriculture and the Interior are about on a par with agricultural scientists. Hence, the salary range for beginners with no prior significant experience is only up to $18,500 per year.

The other information given earlier regarding agricultural scientists would apply, as well, to fisheries scientists.

What Is the Job Future for Fisheries Conservationists?

The American Fisheries Society's current "Careers in Fisheries" folder reports:

The number of available fisheries positions has expanded in recent years as a result of increased funding under the Federal Aid in Sport Fish Restoration Act, which distributes funds collected from a federal tax on fishing tackle and other items to the states for fisheries management programs. The increasing demand for fish in our diets has improved the employment picture for students trained in aquaculture. Graduates with strong educational backgrounds and experience in aquaculture and fisheries research and management are always in demand.

While the current *Occupational Outlook Handbook* does not distinguish fisheries conservationists from other conservationists in its estimates, it would appear that there are more than 16,000 such specialists. Unless there are favorable changes in national priorities, growth can be expected to be modest, perhaps 15 percent, over the next decade, bringing employment to about 19,000 by the year 2005.

FORESTRY

"What are forests?" asks an expert panel in a new report (*Forestry Research,* © 1990 by the National Academy of Sciences, National Academy Press, Washington, D.C.). The report takes its theme from this broad definition:

> Forests and related renewable natural resources include the organisms, soil, water, and air associated with timberlands as well as forest-related rangelands, grasslands, brushlands, wetlands and swamps, alpine lands and tundra, deserts, wildlife habitat, and watersheds. These resources include many different categories of land ownership: national forests, parks, and grasslands; federal, state, and private wildlife and wilderness areas; national, state, county, municipal, and community parks and forests; private nonindustrial timber and range lands; and industrial forests and rangelands.

Forests and their resources are for people to use. Americans used more than 80 cubic feet of wood products per person in 1989, up from 60 cubic feet only 20 years earlier. Worldwide, demand for wood products has nearly doubled in the past three decades. The U.S. Forest Service predicts demand will increase by another 45 percent by the year 2000. Most important for the future, trees also are regarded as the front line of defense against global warming because of their ability to remove huge amounts of carbon dioxide from the atmosphere and "fix" it at the earth's surface.

At the precise time that people expect more from forests and forest research, efforts to learn more actually are being reduced, according to the study committee that wrote the report. Since 1978, the number of undergraduate degrees awarded in forestry and related fields has declined by half. Because of inflation, the static government research budget supports less and less research, while industry's forestry research budget also has decreased.

The report calls for the establishment of a new forestry model, an "environmental" approach to forestry problems. This approach "holds that human beings and nature are interrelated, that humans are not superior to the natural world, but depend on the biosphere for their existence." While lumber production will continue, it must be balanced with other needs and done in such a way as to minimize damage to forest ecosystems.

What Do Foresters Do?

A brochure published by The American Forestry Association, a citizen group, and The Society of American Foresters, a professional group, defines the field of their mutual interest:

Forestry is a science that involves managing forest resources in an increasingly complex world. Forestry is also the profession that must answer that challenge, and therefore requires extensive education and training in science and liberal arts. Thus, a forester is a person educated in the science and art of forestry and engaged in forestry work.

An earlier information sheet is more down to earth in its list of foresters' functions:

They direct land surveys, road construction, and the planting and harvesting of trees, applying the economics of forestry. They are skilled in preventing damage to forest resources from insects, diseases, and fires. They plan and prescribe forestland uses and practices, and work with the people involved. For example, they plan and supervise recreational uses of forestland, timber harvesting crews, fire fighters, and tree planters. Foresters administer forest properties, government agencies, and forest companies. Others research or teach in the field of forestry. Foresters work with the general public and forest owners in making America's forests a world model of health, beauty, and productivity.

What Education Do Foresters Need?

In 1995, 55 colleges and universities offered bachelor's or higher degrees in forestry in programs accredited by the Society of American Foresters.

The professional forester must have at least a bachelor's degree in forestry from one of these schools of forestry. If possible, you should consider the M.S. or Ph.D., especially if you intend to concentrate on teaching and research. The curriculum includes a well-rounded education in the biological, physical, and social sciences. Specialized studies include concentrations in ecology, forest economics, forest protection, silviculture, resources management and use, dendrology, forest measurements, forest policy, and forest administration. Forestry schools usually require the student to spend one summer in a college-operated field camp and encourage spending other summers in related work, if possible.

The environmental science/environmental studies curriculum from a liberal arts college or university might qualify you for a forestry-related job, but probably only at the technician level. You would need the B.S. in forestry for serious consideration as a professional.

Thirteen states have mandatory licensing or registration requirements which a forester must meet in order to acquire the title "Professional Forester." Required, in addition to an acceptable degree, are a minimum period of training and professional practice, and successful completion of an examination.

Where Do Foresters Work?

An estimated 50 percent of professional foresters work for public agencies—federal, state, and local. Some 32 percent are employed by industrial concerns, mainly pulp and paper, lumber, logging, and milling companies. The remaining 18 percent work in the forestry industry, teach, do research, or are in graduate school.

The principal employing organization is the U.S. Forest Service of the Department of Agriculture. The Soil Conservation Service also is a major employer. Many foresters also work in components of the Department of the Interior: Bureau of Land Management, U.S. Fish and Wildlife Service, and National Park Service. A forester employed by a state would work in the comparable state agency.

What Do Foresters Earn?

Most bachelor's graduates entering the federal service as foresters in 1993 started at $18,340 or $22,717 a year, the exact level depending on academic achievement and experience, according to the Bureau of Labor Statistics. Starting salaries in private industry were comparable, while those in state and local government were somewhat lower.

About foresters and conservationists who work for government at any level, the BLS makes an important point: They generally enjoy more generous benefits—pension and retirement plans, health and life insurance, and paid vacations—than do those who work for smaller firms. And the natural resources field has many small firms.

What Is the Job Future for Foresters?

There are an estimated 27,000 professional foresters in the United States. The Bureau of Labor Statistics expects employment of foresters and conservation scientists to grow more slowly than the average for all occupations through the year 2005, partly due to budgetary constraints in the federal government, where employment is concentrated. However, BLS says, an expected wave of retirements in the federal government should create additional job openings for both foresters and range conservationists. Job opportunities for foresters outside of the federal government are expected to be better.

Even though only modest rise is anticipated in the numbers of foresters employed by the year 2005, the BLS finds some hope:

> Demand will continue to increase at the state and local government levels in response to the emphasis on environmental protection and responsible land management. For example, urban foresters are increasingly needed to do environmental impact studies in urban areas, and to help regional planning commissions make land-use decisions, particularly in the Northeast and in the major population centers of the country.

> At the state level, more numerous and complex environmental regulations have created demand for more foresters to deal with these issues. Also, the Stewardship Incentive Program, funded by the federal government, provides money to the states to encourage landowners to practice multiple-use forest management. Foresters will be needed to assist landowners in making decisions about how to manage their forested property. In private industry, more foresters should be needed to improve forest and logging practices and increase output and profitability.

RANGE MANAGEMENT

You have seen it in Western movies on TV: It is native grazing land, called rangelands. It covers about 47 percent of the entire land area of the earth. It is the largest single category of land in the United States— more than one billion acres, mostly in the western states and Alaska.

Rangelands contain many natural resources: grass and shrubs for animal grazing, habitats for livestock and wildlife, facilities for water sports and other kinds of recreation, and areas for scientific study of the environment. These renewable resources can yield their full potential only if properly managed.

What Do Range Managers Do?

The *range manager*—sometimes called *range conservationist, range scientist,* or *range ecologist*—manages, improves, and protects this ecological system. Become a range manager and you would be responsible for deciding the number and kind of animals to be grazed. You would be a practical ecologist, selecting the best season for grazing while conserving soil and vegetation for other uses, such as wildlife grazing, outdoor recreation, watersheds, and growing timber. You would be a practical economist, optimizing the production of livestock and sometimes timber and commercial crops.

You would restore or improve rangelands through techniques such as controlled burning, reseeding, and the biological, chemical, or mechanical control of undesirable plants. You would have the satisfaction of seeing your surroundings change as a result of your work. You might plow up rangelands covered by natural sagebrush vegetation and reseed them with more productive grass.

Because of the multiple use of rangelands, you might spend much of your working time performing the duties of a forester, wildlife conservationist, watershed manager, recreationist, or even a farmer! You might have to provide animal watering facilities, control erosion, and build pens and fences, for example.

What Education Do Range Managers Need?

A bachelor's degree in range management is the normal educational qualification for a professional position in the field. A degree in a closely related field such as agronomy or forestry, including courses in range management, may be accepted. Approximately 40 colleges and universities in the United States, Canada, and Mexico have joined to form the Range Science Education Council. All of the institutions offer course work in range management, but only a few of the programs are accredited by the Society for Range Management. This is a mixture of institutions offering B.S. degrees in a closely related discipline with a range management or range sciences option; some offer bachelor's degrees specifically in those disciplines.

A degree in range management requires a basic knowledge of biology, chemistry, physics, mathematics, and communications skills. Advanced courses combine plant, animal, and soil sciences with principles of ecology and resources management. Desirable electives include economics, computer science, forestry, wildlife, and recreation.

Where Do Range Managers Work?

The majority of range managers work for federal, state, and local governments, almost exclusively in the western half of the United States or Canada. Federal employees work mainly in the Forest Service and Soil Conservation Service of the Department of Agriculture and in the Bureau of Land Management of the Department of the Interior. State and provincial game and fish departments employ many range managers.

In the private sector, livestock ranches are the largest employers. Some range managers work as rangeland appraisers for banks and real estate firms. Others manage their own lands. A few teach and do research at colleges and universities or work overseas with U.S. or United Nations agencies.

What Do Range Managers Earn?

Information given previously in this chapter for other conservation professionals applies to range managers as well.

The Bureau of Land Management regularly advertises for range managers at the GS-5 to GS-7 levels, depending upon education and experience. The starting salary range is $16,875 to $25,569.

What Is the Job Future for Range Managers?

According to the Society for Range Management:

Rangelands contribute more than three billion dollars to the economy of the 17 western states, of which almost two billion dollars are contributed to the 11 "public land" states. The range livestock industry is one of the largest employers in the region and contributes greatly to the rural economy and culture of the West.

Employment opportunities for range managers are expected to grow faster than the average for all occupations, through the year 2000. The growing demands for red meat, wildlife habitat, recreation, and water, as well as increasing environmental concerns, should stimulate the need for more range managers. Because the amount of land cannot be expanded, range managers will need to increase the range ecosystem's productivity while maintaining its environmental quality. Also, range managers will be in greater demand to manage large ranches, which are increasing in number. Some of these "spreads" in the western United States are being acquired by super-rich Middle Eastern oil potentates as investments in a country which they consider safe, prosperous, and inviting. Wealthy Americans, as well, are buying ranches for sport, vacation, and investments.

As oil and coal exploration accelerates, private industry will require many more range specialists to reclaim or restore mine lands to productivity, as required by federal law.

SOIL CONSERVATION

Back in the 1930s, the United States lost millions of acres of fertile, productive soil in the Great Plains states due to drought and wind erosion. Deforestation of vast areas of Africa, Asia, and South America are subjecting the ecology, the resources, and the people to traumatic losses, possibly changing the climates of continents. In most cases, those are crass, preventable losses—and they persist today, on a smaller scale, in this country.

What Do Soil Conservationists Do?

As a *soil conservationist,* you would be responsible for supplying farmers, ranchers, and others with technical assistance. Such landowners are organized under state law into soil conservation districts. While these people would be your "clients," the chances are that you would work for the federal government in helping them to adjust land use, protect land against soil deterioration, rebuild eroded and depleted soils, and stabilize runoff and sediment-producing areas. You also would help improve cover on lands devoted to raising crops and maintaining forests, pasture, and rangelands, and the wildlife they support. In addition, you probably would help plan methods and facilities for handling water for farm and ranch use, conserving water, reducing damage from flood water and sediment, and draining or irrigating farms or ranches as needed.

As a soil conservationist (or *soil scientist* or *soil engineer*) you would draw maps portraying soil, water, vegetation, and structures. You would compile information including cost-benefit analyses (the costs of various uses and treatment of lands and the relative benefits to be expected). You would present these maps and plans to the landowner or operator and answer any questions. When the operator had decided on a plan of action for conservation farming or ranching, you would provide any needed counsel and technical advice and guidance on implementation of the plan.

You might work on a special program of the Soil Conservation Service (SCS), such as snow surveys and water forecasting. Each winter SCS personnel cover about 70,000 miles of the western states by skis, oversnow vehicles, and aircraft to measure the snowpack. The data they collect are translated into a water supply forecast for the following spring and summer planting and growing seasons.

New federal antipollution laws impose strict controls on sediment, or soil runoff, responsible for much water pollution. So, many states employ soil scientists to inspect large highway and building sites where vegetation has been removed and agricultural lands where fertilizers have been applied, to make sure proper erosion control methods were used.

What Education Do Soil Conservationists Need?

As a minimum you should have a bachelor of science degree with a major in soil conservation or a closely related area of natural science or agriculture. Courses in chemistry, physics, mathematics, topography, meteorology, agronomy, and related physical and environmental sciences are desirable. As you would be responsible for communicating and interpreting plans and counseling users on their problems, you should take as many liberal arts courses as your college program permits. These should include English, psychology, speech, and possibly education.

If you were raised on a farm or ranch, you may have an advantage over others in having a "feel" for landowners' needs. If you came from the city or suburbs, on the other hand, you may need to compensate by trying to get summer jobs in rural settings—if necessary, working for room and board to get the experience. The rural culture is quite different from that of the urban environment in which most young persons are raised today.

Where Do Soil Conservationists Work?

Most soil conservationists work for the federal government, mainly the U.S. Department of Agriculture's Soil Conservation Service. Other federal employers are the Forest Service, National Park Service, and Bureau of Reclamation. Private employment is increasing. Banks, public utilities, real estate developers, and consulting engineers also employ soil conservationists.

What Do Soil Conservationists Earn?

The information given previously for agricultural scientists and foresters generally applies to soil conservationists, too.

What Is the Job Future for Soil Conservationists?

The rapid expansion of water supply, wastewater, and other urban environmental programs has accelerated openings for soil conservationists in urban areas, as has industrial development, including the conversion of rural land to industrial parks. Growing demand for new energy sources, for expanding food production, for new housing and new communities, and for reclaiming stripmined lands and the protection of wetlands—all these factors—will contribute to the need for more soil conservationists and related personnel.

There are an estimated 20,000 soil conservationists and scientists in the United States. Despite the need for more, funding constraints will hold growth to about 5 percent, or about 1,000 more positions by the year 2005.

WILDLIFE MANAGEMENT

More than 130 million persons in the United States and Canada, and millions more on all continents, enjoy hunting, viewing, and/or feeding wildlife, according to a survey reported by The Wildlife Society, the organization for professionals. Commercial trapping (and its counterpart, commercial fishing) produce protein nutrition for millions. Farmers and others who hunt feed their families, in part, from what they shoot and bag.

Wild animals are part of the natural heritage of the American and other peoples. Species may need protection when they become scarce. Scarcity or extinction almost always is the result of vital habitats being destroyed or altered by deliberate environmental development. Destruction, sometimes wanton, occurs when people destroy animals such as rabbits, blackbirds, deer, and others that forage on crops, rangelands, or gardens, destroying plants being grown for use.

Wildlife is defined as all animals that are not domesticated. In practical terms, it generally means *game* species that are harvested for food, sport, or other reasons. Game animals are controlled or managed by the manipulation of habitat and by the establishment of seasons, licenses, and bag limits.

What Do Wildlife Managers Do?

Several natural resources professions have moved away from using the word *conservation* in job titles because of its connotations of *protection* or *preservation* rather than beneficial *use* and *regeneration.* (Example: *wildlife conservation* has become *wildlife management.*)

The Wildlife Society now uses the term *wildlife management* and defines it as:

> ...human effort to maintain or manipulate natural resources, including soil, water, plants, and animals (including man) for the best interests of the environment (including man) whether these interests be ecological, commercial, recreational or scientific.

A wildlife manager has a professional goal to assure continued, satisfactory population levels of wildlife. One important job of the wildlife manager is to gather data through research to formulate and to apply scientifically sound solutions to wildlife species or habitat problems. Another job is to conduct on-the-ground management programs, enforce regulations, administer programs or properties. An equally important effort is to inform others about wildlife, its ecology, and its management.

What Education Do Wildlife Managers Need?

You can take suitable bachelor's level work, say, in biology, at almost any college or university, and this would be sufficient for a beginning job in wildlife biology. In addition to a thorough grounding in physical and biological sciences, you would need training in such liberal arts as English, languages, history, geography, statistics, and the economics of food and fiber production. While still in high school, you should take as many mathematics, physics, English, chemistry, and biology courses as you can. The Wildlife Society recommends that you acquire group experience in committee work and meetings and write for high school publications, as well.

If you know upon graduating from high school that wildlife conservation or management is the field for you, you might choose one of the almost 100 colleges and universities listed by The Wildlife Society in its annual survey of enrollments.

The minimum educational requirements for certification are a four-year course of study in an accredited college or university, leading to a bachelor's or higher degree in wildlife. Among the courses required are the biological sciences; wildlife management; ecology and botany; physical sciences; mathematics through calculus; computer science; humanities, social sciences, and English composition; and environmental policy, administration, law, law enforcement, land-use planning, and other electives.

Where Do Wildlife Managers Work?

Most wildlife conservationists in the United States and Canada work for the federal government, state, or provincial governments or in colleges and universities. Many of these programs are cooperative, that is, funded largely by the U.S. government but conducted in state facilities by state employees.

In a 1982 survey, The Wildlife Society found that federal agencies hired the most bachelor's degree graduates, state agencies the most master's degree recipients, and colleges and universities the most doctorate holders. Only university hiring increased over a two-year period.

The federal sector remained the largest employer, hiring 34 percent of those graduates obtaining wildlife employment. Within the federal government, the U.S. Forest Service was the largest overall employer, with 25 percent; it also hired the most bachelor's degree graduates, with the U.S. Fish and Wildlife Service second by both measures.

What Do Wildlife Managers Earn?

The compensation picture for wildlife managers is similar to that of other conservationists covered above.

What Is the Job Future for Wildlife Managers?

There are about 20,000 wildlife managers and conservationists in the United States. A modest 5 percent growth, to 21,000, is expected by the year 2005.

Growth of the wildlife field could be accelerated if some body, such as the National Research Council or the Department of the Interior, were to undertake a study equivalent to *Opportunities in Biology* or *Fisheries Research,* described earlier. Such fact-finding studies often have led to significant revitalization of other fields—for example, environmental protection in the 1970s and 1980s.

WHERE SHOULD YOU WRITE FOR MORE INFORMATION?

American Fisheries Society
    5410 Grosvenor Lane
    Bethesda, Md. 20814-2199

National Wildlife Federation
    1400 Sixteenth Street, N.W.
    Washington, D.C. 20036-2266

Society for Range Management
    1839 York Street
    Denver, Colo. 80206

Society of American Foresters
    5400 Grosvenor Lane
    Bethesda, Md. 20814

Soil Conservation Society of America
    7515 N.W. Ankeny Road
    Ankeny, Iowa 50021

The Wildlife Society
    5410 Grosvenor Lane
    Bethesda, Md. 20814

CHAPTER 6

LAND USE AND HUMAN SETTLEMENTS

Misuse of the land is now one of the most serious and difficult challenges to environmental quality, because it is the most out-of-hand, and irreversible. Land use is still not guided by any agreed upon standards. It is instead influenced by a welter of sometimes competing, overlapping government institutions and programs, private and public attitudes and biases, and distorted economic incentives.—*First Annual Report on Environmental Quality,* 1970.

PROFESSIONS COVERED

Architecture, Geography, Landscape Architecture, Urban and Regional Planning

Social decisions about land use involve the very fabric of society. Individuals and families, constantly changing and frequently moving, create impacts on the land of which they are scarcely aware. So do governmental decisions about development, housing, transportation, agriculture, water resources, and the like. So do private industry's various moves—to build a factory, shopping center, power plant, and so forth. Conflicts of many kinds are inevitable in such drastic social change as land use control. Private property versus public environmental concerns is perhaps the foremost conflict. Economic growth, payrolls, and jobs versus open land and low-density population is another. And environmental protection versus energy needs is a third.

We have conflicting feelings about the land. As we saw in the preceding chapter, we want to preserve much of it for our own and future generations' use and enjoyment. At the same time, we want to be free—and for others to be free—to own and use as much of it as we can afford. Yet, there is a finite amount of land. And the population of the United States still grows by one to two million per year. What kind of environment our children and their children inhabit is being determined today, to some degree, by such public policies as public housing subsidies, mortgage guarantees, and federal tax provisions favoring home ownership.

Land use traditionally has been, and remains, primarily a local responsibility. What one is allowed to do with one's private property is determined by local laws and regulations, established and enforced by local government. Architects, landscape architects, and planners, three of the professionals discussed here, generally work on one project at a time, and it is the local jurisdiction that controls their work there.

TRENDS IN LAND USE

The Council on Environmental Quality has been observing major national trends in land use since its inception in 1970. These include the continuing loss of agricultural lands to urbanization; desertification, the spread of desert-like conditions to areas that once were more productive; population changes, both large geographic shifts westward and southward and suburbanization around cities; and the special problems of the coastal zone and impact of encroachment on wetlands and marine resources. Public lands are being encroached upon at unprecedented rates.

Environmentalists recognized early in the new environmental era that the new tools of social and economic analysis and planning would permit assessing the costs versus the benefits of federal actions affecting the environment. They searched for an "action-forcing" mechanism to assure that the federal government could follow through on any pledges to protect the environment.

The device found and made the keystone of the National Environmental Policy Act (NEPA) of 1969 is its section 102(2)(c). This section requires each federal agency to prepare an environmental impact statement in advance of every major action, recommendation, or report on legislation that may significantly affect the quality of the human environment. Hundreds of different types of activities are covered, and thousands of cases have been filed under NEPA since it was implemented in 1970.

ARCHITECTURE

Architecture is the art and science of design, according to the careers brochure of The American Institute of Architects, which elaborates:

> *Architects* are professionals who organize space in, around, and among buildings to satisfy humans' need for shelter. They must be artists, engineers, social scientists, and environmentalists in their search for creative solutions to design problems that respond to function—how the building "works" for its users; structure—how it will be constructed; form—how effects of volume, shape, color, and texture interact to express the esthetic quality of the whole; economy—how to balance the project requirements within the client's budget; environment—how the building "fits" in its natural and climatic setting; and regulation—how the building protects the safety, comfort, and well-being of its occupants.

Such a broad functional statement barely suggests the range of the architect's work. This work also involves the redesign of existing facilities, the design of new facilities and the preservation and restoration of old, sometimes historic, properties. The architect's work ranges in scale from the design of an individual space to the development of comprehensive urban, regional, even national plans.

What Do Architects Do?

Architects may be engaged in private practice or work for architectural, engineering, and planning firms; for government agencies; commercial, industrial, and institutional organizations such as hotel-motel chains, restaurant operators, or fast-food businesses; and for development and real estate companies.

An individual or firm may choose to specialize in a particular building type: schools, housing, retirement communities, hospitals or nursing homes, even golf courses and other recreation facilities.

Many specialize in such areas as marketing, project management, programming, specifications writing, energy conservation, or estimating and cost control. Others pursue allied professions such as planning, landscape architecture, interior design, or historic preservation.

Architectural services offer programming, the research and analysis of client needs that will form the basis of a design project; schematic design, the preliminary, conceptual stage of a project that fixes its broadest outlines; design development, a "fleshing out" of the scheme with dimensions, materials, and details; construction documents, the preparation of drawings and specifications that fully describe a project for construction; bidding and negotiation, providing assistance to clients in arriving at a contract for construction; and construction administration, overseeing the progress of the work and the payment of contractors.

Architects must be people-oriented; the work requires them to meet frequently with clients to determine needs, advise on decisions, report on progress and costs, and review plans. They consult with other design professionals who may be engaged as members of the project team: structural, mechanical, and electrical engineers, soils and civil engineers; land-use planners and landscape architects; construction cost estimators; lighting, acoustical, and energy conservation specialists; and interior designers—all working under the direction and supervision of the architect, who is responsible for the overall coordination of the work.

These are the professionals who also consult with local planning, zoning, and building code officials, and should be knowledgeable of current codes and regulations affecting site planning and construction. They also may be required to coordinate with neighborhood organizations, lending institutions, developers, and real estate interests, and must be able to work closely with contractors and materials and equipment manufacturers and suppliers.

Architectural firms must expend great effort to market their services to potential clients to ensure a steady and growing workload of projects in the office. This means photography and graphic arts are important professional tools, and firms must produce brochures and proposals that include drawings, photos, and written descriptions of previous work. Increasingly, architects are turning to computer assisted design and drafting (CADD) technology. Everybody has seen CADD's presentations on television, as in demonstrations of how automotive engineers "play with" a model of a car while designing it. This is what architects do with their blueprints and models today. Computer games suggest the potential for manipulation of three-dimensional schematics. Most offices maintain a large library of slides and renderings for presentation purposes.

The architect frequently employs, or works with as consultants, members of the other disciplines covered in this chapter, and even more: builders, land-use lawyers, codes officials, and architectural historians. In recent years, architectural history has emerged as more than an esoteric subject. Fueled by the historic preservation movement, it has become a significant and growing profession in the United States, Canada, Great Britain, and most mainland European countries. The bachelor or master of architectural history is qualification for a professional job in government or the private sector. Every state has an office with historic-preservation responsibilities; one of the federal-state programs administered through that office provides tax credits to citizens, including developers, who restore qualifying properties for renewal and reuse.

What Education Do Architects Need?

All 50 states and the District of Columbia require architects to be licensed to practice. Candidates must successfully complete a rigorous, 4-day examination. Among the areas covered are site and building design, predesign, building systems, structural technology; mechanical, plumbing, electrical, and life safety systems; materials and methods; and construction documents and services. In most states, the first professional degree in architecture must be from one of the approximately 100 accredited schools of architecture. Graduate education beyond the first professional degree may be desirable for research, teaching, and certain specialties. Over half of all architects are from 5-year bachelor of architecture programs intended for students entering from high school. Increasingly, architecture schools are offering 2-year master of architecture programs for those with a preprofessional undergraduate degree in architecture, or a 3-year or 4-year master's program for those with a degree in another discipline.

Graduates of first-professional-degree programs generally must serve an internship under the supervision of a licensed architect before being admitted to the architecture registration examination. Three years of apprenticeship is required in most states.

Where Do Architects Work?

As a new graduate, you usually would begin as a junior draftsperson in an architectural firm. Under supervision, you would make drawings and models of structures. You could advance, after several years, to chief or senior draftsperson responsible for all major details of a set of working drawings, and possibly directing the work of others. If you are outstanding, and if the partners of the firm have an opening, they might make you an associate. Then you would receive, in addition to a salary, a share of the profits. You might, of course, leave your employer and join with others to form your own architectural firm or establish a private practice alone.

Most architects work for architectural firms or for contractors, real-estate and development firms, or other businesses that have large construction programs. Only a small percentage work for the federal government, primarily for the Departments of Defense, the Interior, Housing and Urban Development, and the General Services Administration.

What Do Architects Earn?

According to the American Institute of Architects, graduates of professional degree programs working as intern-architects earn approximately $24,000 a year. Beginning salaries depend upon the size, type, and geographical location of the employing organization.

Those with managerial responsibilities in large architectural firms may have salaries and bonuses exceeding $100,000 a year. The earnings of self-employed architects vary considerably and depend upon the size, nature, and specialization of their offices, the scope of professional services offered, and geographic location.

What Is the Job Future for Architects?

The architectural function is essential to many types of enterprises. Companies as diverse as developers of office buildings and shopping malls, and operators of fast-food chains, hotels/motels, recreation and resort operations, hospitals and nursing homes, all employ architectural staff or contract architects.

All those industries, as well as home building, are vulnerable to economic depression, inflation, and war. In the 1980s, when the price of oil fell due to a glut in supply and falling prices, the economies of Houston, Dallas, and other Sun Belt cities were hard hit, and many skyscrapers and malls remained closed and empty. Concurrently, the savings and loan scandals hit hardest at the building industry. When Iraq's aggression in 1990 called for the United Nations' military buildup in the Persian Gulf, culminating in Operation Desert Storm, the domestic economy was further upset, as always happens in war. The demand for architects is highly dependent upon the local level of construction, residential and nonresidential, as well as upon national economic conditions.

According to the BLS, architects held about 96,000 jobs in the United States in 1992. Most were in architectural firms. About one-third of the total were self-employed, either working in solo practice or as freelancers in other firms. A few worked for builders, real estate developers, or for government agencies such as the Departments of Defense, Interior, and Housing and Urban Development, or for the General Services Administration, the government agency that is responsible for developing and maintaining most federal buildings.

Then there are the significant but logical career changes: from architecture into graphic design, advertising, visual arts, product design, construction contracting and supervision, real estate, or investments. One could get a job with government or with a consulting firm, perhaps preparing environmental impact studies and statements. Teaching, journalism, architecture criticism, photography—all are ancillary areas into which architects can move.

Finally, there always is the possibility of a drastic career change such as joining the Peace Corps or a private philanthropy, say Habitat for Humanity, a nonprofit organization which is building or rehabilitating affordable housing in Appalachia, in New York City, and in the developing nations of Africa and elsewhere.

GEOGRAPHY

The Association of American Geographers, in a careers publication, calls geography "an especially attractive major for Liberal Arts students." Geography provides a foundation for a number of possible occupations, and a regional and world perspective required of responsible citizens.

The most comprehensive of the Association of American Geographers' career guides about the field of geography is the 48-page brochure written by AAG's deputy executive director, Salvatore J. Natoli. In its latest edition, he writes:

> To many individuals geography means knowing the locations of countries, their capital cities, and their major rivers, and learning something about the various occupations and trade relations of different countries of the world. This is an overly simplified view of the field. It does little to describe adequately the scope and depth of geographical knowledge that individuals and organizations use to solve important problems. Thus, to agencies, firms, institutions, and governments that employ geographers, geography means and encompasses a set of skills and a specific point of view that will help them operate efficiently, solve environmental problems, create knowledge, and provide ideas for the future.

What Do Geographers Do?

As a *geographer,* you might use advanced statistical techniques, mathematical models, maps, and a computer. In the field, you might interview people, inspect terrain and features of the "built" environment, and take precise measurements with surveying and meteorological instruments. As a member of a research team, you would have available data and observations made by others: maps, aerial photographs, and data compiled by remote sensing instruments on earth satellites and returned to earth by telecommunications facilities.

The Association of American Geographers emphasizes how practical geography is and cites the geographic studies which first showed how widespread is contamination by radon. Radon is an invisible, odorless radioactive gas produced by the decay of uranium in rock and soil. In 1988, the National Research Council released a report in which it was estimated that radon is responsible for five thousand to twenty thousand lung cancer deaths per year in the United States. As a result, the Environmental Protection Agency inaugurated a program in areas of highest incidence to shield children in schools, and the public in federal buildings, from radon emissions. This was followed by a nationwide public information program to alert parents, in particular, to the hazards of overexposure of children to radon gas. Test kits for the consumer are quite inexpensive, and if a building is found to have a high radon level, steps can be taken to correct the problem.

Most geographers specialize. *Economic geographers* focus on the geographic distribution of economic activities. *Political geographers, urban geographers,* and *regional geographers* help define political boundaries and plan governmental activities. *Physical geographers* study the impacts of the earth's configuration on human and environmental activities. *Cartographers* compile and interpret data and design and construct maps and charts. *Medical geographers* study the effects of the environment on human health, working with environmental health officials, biostatisticians, and others to project disease trends and to plan epidemiological control methods and health care facilities. Those major specialties within geography can be broken down further into many subfields.

In 1985 the National Geographic Society, publisher of the *National Geographic* magazine, launched a long-term, nationwide campaign to improve the quality of geography instruction in elementary and secondary schools. The society had ample evidence, from studies and polls, that young Americans were geographically illiterate. Many could not look at a world map and find the Persian Gulf, Vietnam, or the Soviet Union. Many misnamed the nation to the immediate south of the United States or could not distinguish Iowa from Idaho on the U.S. map.

In an international survey in 1989, young Soviet adults scored significantly higher than their American counterparts in identifying locations on a world map. The 18- to 24-year-olds in the Soviet Union correctly identified an average of 9.3 out of 16 locations worldwide. So did young adults surveyed in Canada and Italy, and Swedish students scored even higher. But in the United States, the average, in 16 tries, was only 6.9 correct answers.

What Education Do Geographers Need?

Approximately 300 U.S. and Canadian colleges and universities offer, at 375 campuses, a 4-year undergraduate degree with a geography concentration. Required are a minimum of 24 semester hours or 36 quarter hours beyond introductory geography courses. Degree requirements vary among colleges, but most require a broad base of course work which may include physical, human, cultural, economic, regional geography; meteorology and climatology; cartography, map interpretation, and map design; remote sensing and air photo interpretation; geographic information systems; environmental studies or resource management; and tourism or planning.

"What can you expect in a course of study leading to a bachelor's degree in geography?" asks AAG director Natoli. Rhetorically, he answers:

> The most important part of your career training for geography or for any other career field will be the liberal arts background you obtain in college, including the liberal arts requirements in geography. Without this rich background, many of the specialized courses you take later will have little meaning.

Although many graduates obtain positions with a bachelor's degree, better-paying, more challenging positions usually require at least a master's degree.

Where Do Geographers Work?

About one-third of professional geographers teach in colleges and universities. The federal government is the second-largest employer, principally in the Departments of Defense, Interior, Commerce, and Agriculture. Other agencies with significant cadres of geographers are the Environmental Protection Agency, Smithsonian Institution, National Aeronautics and Space Administration, and the Central Intelligence Agency. State and local governments employ others, in the agencies comparable to those named above for the federal government.

Magazine and book publishers such as the National Geographic Society employ some geographers, as do the publishers who produce maps, including the oil companies' road maps, geologists' maps, and atlases. Increasingly, consulting engineering and management firms hire geographers.

What Do Geographers Earn?

According to the Bureau of Labor Statistics, bachelor's level geographers, when they are hired by the federal government, start at the GS-5 level, which paid approximately $18,500 per annum in 1995. Starting salaries in private industry would be slightly higher, and in state and local government somewhat lower. Colleges and universities require master's degrees for faculty appointments, but graduate fellowships are sometimes available.

What Is the Job Future for Geographers?

In a recent search of jobs newsletters, no jobs were found for geographers per se. However, numerous openings apparently were broad enough for such an education to be qualifying. These were jobs, among others, as analyst, ombudsman, trainer, planner, information and education specialist, teacher, technical publications editor, school or camp director, and environmental educator. They called, generally, for "degrees or experience in environmental sciences or related field."

The Bureau of Labor Statistics sees little or no employment growth in colleges and universities. But it is somewhat optimistic about opportunities in urban and environmental management and planning, including location analysis, land and water resources management, and environmental health planning. "Those with strong backgrounds in cartography, remote sensing imagery interpretation, computer mapping, physical geography, and quantitative techniques should be in particular demand," says the BLS, adding:

> The federal government will need additional personnel to work in programs such as health planning, regional development, environmental quality, and intelligence. Employment of geographers in state and local government is expected to expand, particularly in health planning; conservation; environmental quality; highway planning; and city, community, and regional planning and development. Private industry is expected to hire increasing numbers of geographers for market research and location analysis.

LANDSCAPE ARCHITECTURE

At least two factors go into the achievement of an attractive new or restored community, office or factory building, shopping center, airport, college campus, or, for that matter, residence. Foremost is the building or buildings. Second is the setting or surroundings, combined with the landscaping. Even an ordinary brick or glass-wall low-rise office building, warehouse, or factory can be appreciated fully if the trees, shrubbery plantings, flower beds, and lawns are tastefully designed and the design skillfully executed. The professional often contracted with to design these adornments—to "make it all work together"—is the *landscape architect.*

What Do Landscape Architects Do?

Some nursery companies offer landscaping services, free with purchase of nursery stock or on payment of a fee, to the average individual homeowner. Owners of large estates may employ professional landscape architects to establish initial landscape plans, to maintain the developed landscape, or to make significant changes on the premises. Otherwise, except perhaps in the most affluent communities, such professionals work mainly on public, commercial, or multiunit residential projects.

Landscape architects are employed by real estate development firms starting new industrial-park projects, municipalities constructing airports or public parks, or investors planning hotel, golf course, or similar resort developments. They often are involved in the project from conception, participating with investors, bankers, lawyers, planners, zoning officials, and historic preservationists in assessing the suitability of, and adapting, the property for the intended purpose. The owners-to-be may ask these professionals' advice on the feasibility of spending the necessary investment versus the expected payoff. A big development project requires complicated teamwork.

Once the project is begun, landscape architects work closely with architects and engineers to help determine the best arrangement of roads and buildings. They create detailed plans indicating new topography; vegetation such as trees or shrubbery to be retained; trees, shrubs, and flower beds to be added; walkways, driveways, parking areas; such recreation amenities as tennis courts and ponds, if any; artificial lighting; and inviting entranceways to facilities. The service entrance must be made accessible but inconspicuous and the refuse collection area shielded from public view.

Often-subtle factors include the natural elements of the site, such as the climate, soil, rock outcroppings, vegetation, slope of the land, and drainage. The designers must consider open spaces and anticipate the patterns cast upon the building by sunlight, clouds, and other images at different seasons. If there is a stream or lake, a historic or an archaeological site, a cemetery or Native American burial ground—all these must be considered for preservation.

If desirable and appropriate, resting places, picnic tables, benches, a garden, or bicycle and bridle trails, would be incorporated early.

After studying and analyzing the site and its amenities, the landscape architect prepares a preliminary design, which must be submitted to the client for approval. Many changes usually must be made before a design is accepted. New techniques of computer-aided design (CADD) systems make the design stages much faster and far more cost effective than ever before.

The landscape architect makes lists of building materials, determines specifications, and invites landscape contractors to bid for the work.

Costs are always a factor in a building project and tend to rise because the prices of components are constantly rising—and because clients change their minds. The landscape architect is responsible, with others on the design team, for preparing careful cost estimates, economizing wherever possible, revising regularly, and keeping the client and team colleagues apprised of any changes. The actual design work, assisted by computer programs, involves sketches, models, written reports, photographs, land-use slides, and never-ending cost changes.

Some landscape architects specialize in a particular type of work, such as residential development, historic landscape restoration, waterfront improvements, parks and playgrounds, shopping centers, or industrial parks. Still others work in regional planning and resource management, urban redevelopment, environmental impact assessment, or energy conservation through plantings; a properly sited or shaded building can be heated or cooled with a minimum of fuel consumption.

What Education Do Landscape Architects Need?

A bachelor's degree in landscape architecture, which takes 4 or 5 years, is the minimum educational requirement for entering the profession. About 50 colleges offer programs accredited by the American Society of Landscape Architects.

Most schools advise the high school student to take courses in art, botany, and more mathematics than the minimum required for college entrance. Some also require a high school course in mechanical or geometrical drawing. Otherwise, the requirements for admission are similar to those for architecture, engineering, or science. College technical courses would include surveying, landscape and architectural design, landscape construction, plant materials and design, sketching, recreation and city planning, contracts, specifications, cost estimates, and business

practices. You should especially enjoy the field trips, and if you are for-
tunate, you may get some practical job experience while still in college.

Forty-four states require a license, based on the results of a uniform
national licensing examination, for independent practice of landscape
architecture.

Where Do Landscape Architects Work?

Three-fifths of all landscape architects are in the private sector, working in their own, usually small, businesses, or for architectural, landscaping, or engineering firms. Others are employed by government
agencies concerned with forest management, water storage, public
housing, city planning, urban renewal, highways, parks, recreation, or
energy conservation. Of the total 19,000 licensed landscape architects,
fewer than 1,000 work for the federal government, mainly in the Departments of Agriculture, Defense, Energy, and Interior.

What Do Landscape Architects Earn?

Refer to the section on earnings for architects. Even though landscape
architecture is a different field, it is practiced in much the same manner,
and earnings scales for one approximate those for the other.

What Is the Job Future for Landscape Architects?

The Bureau of Labor Statistics expects employment of landscape architects to increase about as fast as the average for all occupations
through the year 2005. The level of new construction and land development plays an important role in determining demand for landscape architecture. The BLS expects anticipated growth in construction to
increase demand for these services over the long run.

The BLS's *Occupational Outlook Handbook* foresees good news and
bad news:

An increasing proportion of office and other commercial and
industrial developments will occur outside cities. These projects

are typically located on larger sites with more surrounding land which needs to be designed, in contrast to urban development, which often includes little or no surrounding land. Also, as the cost of land increases, the importance of good site planning and landscape design increases. Because employment is linked to new construction, however, landscape architects may face layoffs and competition for jobs when real estate sales and construction slow down, such as during a recession.

URBAN AND REGIONAL PLANNING

The term *urban and regional planning* is used here, and by the Bureau of Labor Statistics (BLS) and the American Planning Association (APA), to distinguish this type of planning from social planning or corporate and business planning. Both types of planners belong to the professional association, the APA.

Urban and regional planning, in the definition of the APA, "is a systematic, creative approach used to address and resolve social, physical, and economic problems of neighborhoods, cities, suburbs, metropolitan areas, and larger regions." This planning is focused on the human environment. Its practitioners might be called community or city planners. We choose to call them urban and regional planners.

What Do Urban and Regional Planners Do?

Urban and regional planners, according to the BLS, develop programs to provide for growth and revitalization of urban, suburban, and rural communities and their regions. Planners address such issues as central city redevelopment, traffic congestion, and the impact of growth and change on an area. They formulate capital improvement plans to construct new school buildings, parks and playgrounds, public housing, and water and sewage systems. They explore locations for sewage treatment and solid-waste disposal facilities, and for chemical, hazardous, and nuclear waste sites. They develop policies and programs for the protection of wetlands, wildlife refuges, archaeological sites, and historic monuments. As the last of the unprotected Civil War battlefields is threatened by development, planners are searching for ways the land can be bought and the landscape preserved.

Inquire about careers and the APA will send you a packet of materials elaborating on the above description, with a letter which says:

> Planning is a profession in which a diversity of techniques is utilized to bring about positive change. As such, planners must bring a broad perspective to problem-solving efforts affecting a range of social, economic, environmental, and political concerns. The training that planners receive serves them well in many types of employment situations.

At times, planners work with every other environmental manager discussed in this book, most frequently with architects, geographers, and landscape architects. And they work with government officials and politicians.

Planners increasingly are becoming involved in social issues such as the housing needs of minorities and diverse cultures, an aging population, housing and treatment facilities for people with AIDS, shelters for the homeless, and drug treatment centers. Planners work closely with those of the ancillary professions, including—in the cases just cited—social workers, physicians, elective officials, city managers, lawyers, and civic activists and advocates, such as those working with the homeless and people with AIDS.

How frequently planners meet such special circumstances depends on where they work, and it would be most frequent in urban areas. More commonly, planners spend their time examining universal community needs such as for schools, libraries, and health clinics, and the necessities and amenities of modern life: office buildings and industrial parks, shopping malls, transportation, roads, parking facilities, commuting travel patterns and loads, historic preservation and restoration, downtown restoration and reuse, parks and recreation areas, and pathways and bicycle paths. They must consider economic and legal issues and the environmental impacts of such development.

Inherent in much environmental change is conflict, along with social considerations and economic consequences—usually, somebody loses and somebody wins. The planner is in the middle of some community disputes and often must facilitate resolution of the conflict. This can either be accomplished informally or by more formal conciliation, arbitration, or mediation.

Environmental mediation is an important new adjunct profession in environmental management, and the planner frequently plays a key role as a mediator. The planner also may have to testify before hearing boards and political bodies, participate in public forums and community meetings, and give interviews to radio, television, and newspaper journalists.

The range of such conflicts and the challenge to the planner are suggested by a recent issue of the APA monthly journal *Planning*. It features articles on how the U.S. Department of Housing and Urban Development is responding to the disastrous national housing fraud scandal that was uncovered there in the late 1980s, annexation battles, the planning profession's code of ethics, and environmental law, as well as how public bodies are acquiring inner-city religious buildings (when the congregations move to the suburbs and want to sell them, or their members grow old and the congregations disappear), preserving, renovating, and converting them to everything from housing to community centers to theaters.

What Education Do Planners Need?

There are ten colleges and universities that offer an accredited bachelor's degree program in urban or regional planning, and about 80 offering a master's. Increasingly, the master's is required for professional appointments. These programs are accepted by the Planning Accreditation Board, sponsored by the American Institute of Certified Planners and the Association of Collegiate Schools of Planning, both of which are affiliated with APA. Electives generally include demography, economics, finance, health administration, location theory, and management.

The bachelor's in planning, or in architecture, architectural history, engineering, or environmental studies, may qualify one for a beginning position in urban and regional planning. Often, work toward the master's can be pursued part-time. Certification is provided to individuals who have the appropriate combination of education with the master's and professional experience.

Where Do Planners Work?

Most planners work for city, county, or regional planning agencies. The number of professionals in private business and research organizations is increasing, however. Planners also work for banks, other savings institutions, airline companies, utilities, and other firms that provide services to urban and rural areas, especially when they are subject to governmental regulation. According to the BLS, urban and regional planners held 28,000 jobs in 1992.

While the planning field, more and more, is stressing professionalism, still it is general enough to provide many entry-level jobs for environmentalists who might find a general environmental studies education qualifying.

Overseas, planners work for the Agency for International Development (AID), the World Bank, engineering and consulting firms, and the governments of developing nations.

What Do Planners Earn?

Bachelor's degree holders earned a median salary of $39,200, according to a 1991 survey by the American Planning Association. Included in the study were experienced practitioners, employees of firms large and small, and business owners.

A better measure might be the BLS's report that, in 1993, planners with a master's degree were hired by the federal government at an average starting salary of $27,800. In some cases, persons having fewer than 2 years of graduate work could enter federal service as interns at yearly salaries between $18,300 and $22,700.

Every week metropolitan newspapers such as the *Washington Post,* the *New York Times,* and the *Los Angeles Times* carry classified advertising for planners, usually specifying a specialty: environmental/natural resources, neighborhood, parks, trails, or transportation. All the salary ranges mentioned above are included—from hourly rates as low as $5 for summer and temporary work, to responsible management positions at $60,000 per annum and greater.

What Is the Job Future for Planners?

Employment of urban and regional planners is expected to grow about as fast as the average for all occupations through the year 2005, according to the Bureau of Labor Statistics. BLS tells why:

The continuing importance of transportation, environmental, housing, economic, and energy production planning will spur demand for urban and regional planners. Specific factors contributing to job growth include commercial development to support suburban areas with rapidly growing populations: legislation related to the environment, transportation, housing, and land use and development, such as the Clean Air Act; historic preservation and rehabilitation activities; central city redevelopment; the need to replace the nation's infrastructure, including bridges, highways, and sewers; and interest in zoning and land-use planning in undeveloped and nonmetropolitan areas, including coastal and agricultural areas.

WHERE SHOULD YOU WRITE FOR MORE INFORMATION?

The American Institute of Architects
    Director, Careers in Architecture Program
    1735 New York Avenue, N.W.
    Washington, D.C. 20006

American Planning Association
    1776 Massachusetts Avenue, N.W.
    Washington, D.C. 20036

American Society of Landscape Architects
   4401 Connecticut Avenue, N.W.
   Washington, D.C. 20008-2302

Association of American Geographers
   1710 16th Street, N.W.
   Washington, D.C. 20009-3198

National Geographic Society
   (Publication: *National Geographic*)
   Geography Education Program
   17th and M Streets, N.W.
   Washington, D.C. 20036

CHAPTER 7

TOMORROW'S NEW
ENVIRONMENTALISM

> Our commitment to a sustainable future begins with a strong
> conservation ethic among our young persons, an ethic which must
> be nurtured by example and furthered through education. Thus
> committed, we can look forward to the twenty-first century as the
> Century of the Environment—Dr. Jay Hair, President, National
> Wildlife Federation.

Even though the future will not be a replication of the past, some features of the future are highly predictable. We can, with confidence, pre-
dict the college-age or the work-force population in any future year.
And we know that the U. S. population is growing, and how fast.

On the first Earth Day in April 1970, according to the Bureau of the
Census, the United States had a population of 203,302,031 (which
makes it astonishing that an estimated 10 percent of all Americans participated in that single Earth Day's citizen activities!). By 1995, the population had reached 263,434,000.

The late 1990s and the early part of the next century will be affected
by the wave of children born to the baby boomers. By 2005, the Census
Bureau (using certain assumptions called "the middle series") projects
the population to be 288,286,000. In the year 2010—and for the first
time—the United States will exceed (by 431,000) a population of
300,000,000.

Look at how the Bureau of Labor Statistics (BLS) analyzes the census
figures and makes its work-force projections: until the year 2005 (the
statistics show, in summary) the labor force will grow at about the rate
of recent decades; by then, nearly all of the baby boomers will be over

45 years old, and some will be over 55 and retiring. Because this age group outnumbers younger workers, the average age of the labor force will rise.

The children of the baby boomers will pass from elementary school to high school to college, and into the work force. Women's share of the work force will continue its long-term increase, but the gap between women's faster employment growth rates, compared to men's, will narrow significantly, according to the BLS.

On average, employment will grow faster in the major occupational groups, such as the professions, that require the most education and training, according to the BLS. Jobs in these occupations also offer higher than average earnings. Health and education dominate the lists of occupations forecast to grow the most.

Environmental management may rank with health and education to form "the big three" among the occupational groups with the most favorable employment prospects.

The Ecological Society of America, in its careers brochure, observes: "The number of environmentally-oriented jobs is expected to increase in the next few years....The economy and the environment will affect the number of jobs, with a poor economy decreasing opportunities and a deteriorating environment increasing opportunities."

As noted in the preface to this book, during the first generation of the environmental era (roughly from 1970 to 1990) professional jobs in environmental management more than doubled, from half a million to more than a million. Depending upon how one delineates the field of environmental management, how one defines professionals, and how one calculates employment of professionals, we could be describing a field that today employs two to three million professionals. We call these professionals the *new environmentalists*. Certainly, a doubling every two to three decades could not continue. So, jobs are not as plentiful as they once were—which is true in most fields.

With any large, established field, the majority of job openings are created by the need to replace workers who retire, die, or change jobs. It is because of this job replacement and turnover, rather than more or less federal funding or programs, that environmental management is a large field with extensive work force turnover.

The Bureau of Labor Statistics, as Dr. Hair of the National Wildlife Federation notes in the foreword, "projects a healthy demand into the twenty-first century for well-educated personnel in many areas of environmental management."

The challenge to every young person is to choose an area and become well prepared in that field. If you are that young person...if you are highly motivated to improve your environment...if you are people-oriented...if you are intelligent and disciplined...and if you work hard to prepare yourself to make your full contribution in life, you should find bright prospects in an environmental career.

CITIZEN ENVIRONMENTAL ORGANIZATIONS

(Not Previously Listed)

American Forests, 1516 P Street, N.W., Washington, D.C. 20005. Publication: *American Forests.*

The American Museum of Natural History, Central Park West at 79th Street, New York, N.Y. 10024. Publication: *Natural History Magazine.*

The Cousteau Society, 930 West 21st Street, Norfolk, Va. 23517. Publication: *Calypso Log.*

Defenders of Wildlife, 1101 14th Street, N.W., Washington, D.C. 20005. Publication: *Defenders.*

The Elm Research Institute, Harrisville, N.H. 03450. Publication: *Elm Leaves.*

Environmental Action, 6930 Carroll Avenue, Tacoma Park, Md. 20903. Publication: *Environmental Action.*

Environmental Defense Fund, 444 Park Avenue South, New York, N.Y. 10016. Publication: *EDF Letter.*

Friends of the Earth, 1045 Sansome Street, San Francisco, Calif. 94111. Publication: *Not Man Apart.*

Greenpeace USA, 1436 U Street, N.W., Washington, D.C. 20009. Publication: *Greenpeace Magazine.*

The Izaak Walton League of America, 1401 Wilson Boulevard, Arlington, Va. 22209. Publication: *Outdoor America.*

League of Conservation Voters, 1707 L Street, N.W., Washington, D.C. 20036.

National Audubon Society, 950 Third Avenue, New York, N.Y. 10022. Publication: *Audubon.*

National Parks & Conservation Association, 1776 Massachusetts Avenue, N.W., Washington, D.C. 20036. Publication: *National Parks & Conservation Magazine.*

National Trust for Historic Preservation, 1785 Massachusetts Avenue, N.W., Washington, D.C. 20036. Publication: *Historic Preservation.*

Natural Resources Defense Council, 122 East 42nd Street, New York, N.Y. 10017. Publication: *The Amicus Journal.*

The Nature Conservancy, 1815 North Lynn Street, Arlington, Va. 22209. Publication: *Nature Conservancy.*

Rachel Carson Council, 8940 Jones Mill Road, Washington, D.C. 20815.

Sierra Club, 730 Polk Street, San Francisco, Calif. 94109. Publication: *Sierra.*

Smithsonian Institution, 1000 Jefferson Drive, S. W., Washington, D.C. 20560. Publication: *Smithsonian.*

The Wilderness Society, 900 17th Street, N.W., Washington, D.C. 20006. Publication: *The Living Wilderness.*

Zero Population Growth, 1400 16th Street, N.W., Washington, D.C. 20036. Publication: *The ZPG Reporter.*

APPENDIX B

ENVIRONMENTAL PUBLICATIONS

(Not Previously Listed)

ChemEcology, Chemical Manufacturers Association, 2501 M Street, N.W., Washington, D.C. 20037.

Common Ground, The Conservation Fund, 1800 North Kent Street, Suite 1120, Arlington, Va. 22209.

E: The Environmental Magazine, 26 Knight Street, Norwalk, Conn. 06851.

The Earth Times, Box 3363, Grand Central Station, New York, N.Y. 10163.

Environment, 4000 Albemarle Street, N.W., Suite 504, Washington, D.C. 20016.

Environment Writer, Environmental Health Center, The National Safety Council, 1019 19th Street, N.W., Suite 401, Washington, D.C. 20036.

Resolve, Center for Environmental Dispute Resolution, Resolve, Inc., 2828 Pennsylvania Avenue, N.W., Suite 402, Washington, D.C. 20009.

SEJ Journal, Society of Environmental Journalists, P. O. Box 27506, Philadelphia, Pa. 19118.

APPENDIX C

KEY FEDERAL AGENCIES

Department of Agriculture, Washington, D.C. 20250
  Forest Service; Science and Education Administration; Soil Conservation
  Service.
Canadian Heritage, Hull, Quebec, Canada, K1A 0M5.
Department of Commerce, Washington, D.C. 20230
  National Oceanic and Atmospheric Administration, Rockville, Md. 20852.
Congress of the United States, Washington, D.C. 20510
  Office of Technology Assessment.
Consumer Product Safety Commission, Washington, D.C. 20207.
Council on Environmental Quality, Washington, D.C. 20006.
Department of Defense, Washington, D.C. 20301.
Department of Education, Washington, D.C. 20202.
Department of Energy, Washington, D.C. 20585.
Environment Canada, 10 Wellington Street, Hull, Quebec, Canada K1A 0H3.
Environmental Protection Agency, Washington, D.C. 20460
  Environmental Research Center, National Training and Operational
  Technology Center, Cincinnati, Ohio 45268
  Environmental Research Laboratory, Research Triangle Park, N.C. 22709.
Federal Emergency Management Agency, Washington, D.C. 20472.
Health Canada, Tunney's Pasture, Ottawa, Ontario, Canada K1A 0K9.
Department of Health and Human Services, Washington, D.C. 20201
  Centers for Disease Control and Prevention, Atlanta, Ga. 30333
  National Institute for Occupational Safety and Health, Rockville, Md. 20857
  Food and Drug Administration, Rockville, Md. 20892
  National Institutes of Health, Bethesda, Md. 20205
  National Institute of Environmental Health Sciences, Research Triangle Park,
  N.C. 22709.
Department of Housing and Urban Development, Washington, D.C. 20410.

Department of the Interior, Washington, D.C. 20240
  Fish and Wildlife Service
  National Park Service.
Department of Labor, Washington, D.C. 20210
  Bureau of Labor Statistics; Occupational Safety and Health Administration.
National Aeronautics and Space Administration, Washington, D.C. 20546.
National Science Foundation, Washington, D.C. 20550.
Natural Resources Canada, Place Vincent Massey, 351 St. Joseph Blvd., Hull,
  Quebec, Canada K1A 1G5.
Smithsonian Institution, Washington, D.C. 20560.
Department of Transportation, Washington, D.C. 20590.
Veterans Administration, Washington, D.C. 20420.

A complete list of titles in our extensive *Opportunities* series

OPPORTUNITIES IN

Accounting
Acting
Advertising
Aerospace
Airline
Animal & Pet Care
Architecture
Automotive Service
Banking
Beauty Culture
Biological Sciences
Biotechnology
Broadcasting
Building Construction Trades
Business Communication
Business Management
Cable Television
CAD/CAM
Carpentry
Chemistry
Child Care
Chiropractic
Civil Engineering
Cleaning Service
Commercial Art & Graphic
   Design
Computer Maintenance
Computer Science
Counseling & Development
Crafts
Culinary
Customer Service
Data Processing
Dental Care
Desktop Publishing
Direct Marketing
Drafting
Electrical Trades
Electronics
Energy
Engineering
Engineering Technology
Environmental
Eye Care
Farming and Agriculture
Fashion
Fast Food
Federal Government
Film
Financial

Fire Protection Services
Fitness
Food Services
Foreign Language
Forestry
Franchising
Gerontology & Aging Services
Health & Medical
Heating, Ventilation, Air
   Conditioning, and
   Refrigeration
High Tech
Home Economics
Homecare Services
Horticulture
Hospital Administration
Hotel & Motel Management
Human Resource Management
Information Systems
Installation & Repair
Insurance
Interior Design & Decorating
International Business
Journalism
Laser Technology
Law
Law Enforcement & Criminal
   Justice
Library & Information Science
Machine Trades
Marine & Maritime
Marketing
Masonry
Medical Imaging
Medical Technology
Mental Health
Metalworking
Military
Modeling
Music
Nonprofit Organizations
Nursing
Nutrition
Occupational Therapy
Office Occupations
Paralegal
Paramedical
Part-time & Summer Jobs
Performing Arts
Petroleum
Pharmacy
Photography

Physical Therapy
Physician
Physician Assistant
Plastics
Plumbing & Pipe Fitting
Postal Service
Printing
Property Management
Psychology
Public Health
Public Relations
Publishing
Purchasing
Real Estate
Recreation & Leisure
Religious Service
Restaurant
Retailing
Robotics
Sales
Secretarial
Social Science
Social Work
Special Education
Speech-Language Patholog
Sports & Athletics
Sports Medicine
State & Local Government
Teaching
Teaching English to Speake
   of Other Languages
Technical Writing &
   Communications
Telecommunications
Telemarketing
Television & Video
Theatrical Design &
   Production
Tool & Die
Transportation
Travel
Trucking
Veterinary Medicine
Visual Arts
Vocational & Technical
Warehousing
Waste Management
Welding
Word Processing
Writing
Your Own Service Business

VGM Career Horizons
a division of *NTC Publishing Group*
4255 West Touhy Avenue
Lincolnwood, Illinois 60646–1975